The PyRosetta Interactive Platform for Protein Structure Prediction and Design

A Set of Educational Modules
Second Edition – Talaris Update

Jeffrey J. Gray
Sidhartha Chaudhury
Sergey Lyskov
Jason W. Labonte

Chemical & Biomolecular Engineering
Program in Molecular Biophysics
Johns Hopkins University
Baltimore, Maryland

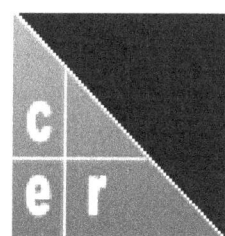

Copyright © 2009 Jeffrey J. Gray, Sidhartha Chaudhury, and Sergey Lyskov
Copyright © 2012, 2014 Jeffrey J. Gray, Sidhartha Chaudhury, Sergey Lyskov, and Jason W. Labonte
Revisions by Evan H. Baugh and Alex J. Mathews
ISBN-13 978-1500968274
Baltimore, Maryland

Visit the PyRosetta web site: http://www.pyrosetta.org

Table of Contents

Preface 1
Workshop #1: PyMOL 3
    Suggested Readings 3
    Basic Operations 3
    Structural Analysis 5
    Comparing Molecules 6
    High-Quality Visualization and Scripting 7
Workshop #2: PyRosetta 9
    Basic Elements 9
    Basic Python 10
    Basic PyRosetta 10
    Protein Geometry 12
    Manipulating Protein Geometry 13
    Visualization and the PyMOL Mover 14
    Programming 14
    Programming Exercises 15
    References 16
Workshop #3: PyRosetta Scoring 17
    Scoring Poses 17
    Energies and the PyMOL Mover 21
    Programming Exercises 21
    References 22
Workshop #4: PyRosetta Folding 23
    Suggested Readings 23
    A Simple *de Novo* Folding Algorithm 23
    Low-Resolution (Centroid) Scoring 24
    Protein Fragments 26
    Programming Exercises 27
    Thought Questions 27
Workshop #5: PyRosetta Refinement 29
    Suggested Reading 29
    Introduction 29
    Small and Shear Moves 30
    Minimization Moves 31
    Monte Carlo Object 32
    Trial Mover 33
    Sequence and Repeat Movers 34
    Refinement Protocol 35
    Programming Exercises 35
    Thought Questions 36
Workshop #6: Packing & Design 37
    Suggested readings 37

- Side Chain Conformations, the Rotamer Library, and Dunbrack Energies 37
- Monte Carlo Side-Chain Packing 38
- Packing for Refinement 38
- Design 39
- Programming Exercises 41
- Thought Question 42
- References 43

Workshop #7: Docking 45
- Suggested Readings 45
- Fast Fourier Transform Based Docking via ZDOCK 45
- Docking Moves in Rosetta 45
- Low-Resolution Docking via RosettaDock 48
- Job Distributor 49
- High-Resolution Docking 50
- Docking Funnel 51
- Programming Exercises 52

Workshop #8: Loop Modeling 53
- Suggested Readings 53
- Fold Tree 53
- Cyclic Coordination Descent (CCD) Loop Closure 56
- Multiple Loops 56
- Loop Building 56
- High-Resolution Loop Protocol 57
- Kinematic Closure (KIC) Protocols 57
- Simultaneous Loop Modeling and Docking 58

Workshop #9: Custom Movers & Energy Methods 60
- Suggested Reading 60
- Classes in Python 60
- Custom Mover Classes 63
- Decorators in Python 64
- Custom Energy Methods 65
- Programming Exercises 68

Coda 71

Appendix A: Command Reference 73

Appendix B: Residue Parameter Files 85

Appendix C: Cleaning pdb Files 89

Appendix D: Links to Online Help 91
- PyMOL 91
- Python 91
- PyRosetta 91
- Rosetta 91

Preface

Structures of proteins and protein complexes help explain biomolecular function. Computational methods provide an inexpensive way to predict unknown structures, manipulate behavior, and design new proteins or functions. The protein structure prediction program Rosetta, developed by a consortium of laboratories in the Rosetta Commons, has an unmatched variety of functionalities and is one of the most accurate protein structure prediction and design approaches (Das & Baker, *Ann Rev Biochem* 2008; Gray, *Curr Op Struct Biol* 2006). To make the Rosetta approaches broadly accessible to biologists and biomolecular engineers with varied backgrounds, we developed PyRosetta, a Python-based interactive platform for accessing the objects and algorithms within the Rosetta protein structure prediction suite. In PyRosetta, users can measure and manipulate protein conformations, calculate energies in low- and high-resolution representations, fold proteins from sequence, model variable regions of proteins (loops), dock proteins or small molecules, and design protein sequences. Furthermore, with access to the primary Rosetta optimization objects, users can build custom protocols for operations tailored to particular biomolecular applications. Since the program can communicate with the visualization software PyMOL, search algorithms can be viewed on-screen in real time.

In this book, we have compiled a set of workshops to teach both the fundamentals and the practical application of protein structure prediction and design. The workshops assume basic knowledge of protein structure and familiarity with computers and the Python programming language. Readings and references are provided in each chapter for more in-depth study. Each workshop covers a single topic in the field and walks the reader through the basic operations in a one- to two-hour session. Interactive exercises are incorporated so that the reader gains hands-on experience using the variety of commands available in the toolkit. The text is arranged progressively, beginning with an introduction to the PyMOL visualization package, proceeding through the fundamentals of protein structure and energetics, and then progressing through the applications of protein folding, refinement, packing, design, docking, and loop modeling. A set of tables is provided at the end of the book as a reference of the available commands.

Additional resources on the Rosetta program are available online. The PyRosetta web site, pyrosetta.org, includes additional example and application scripts. At the web-based user forum www.rosettacommons.org/forum, the PyRosetta community shares questions, answers, and useful scripts. Documentation on Rosetta in general including expert details on the underlying C++ code is available at www.rosettacommons.org/support. PyRosetta is built upon the Rosetta 3 platform, so objects available in PyRosetta will have the same underlying data structures and functionality.

The bulk of these modules were created at the Homewood campus of Johns Hopkins University over the course of two semesters, Spring 2008 and Spring 2009, for the Chemical & Biomolecular Engineering class "Computational Protein Structure Prediction and Design," and further developed in 2010, 2012, and 2014. We acknowledge the contributions of the many developers of the Rosetta community (see www.rosettacommons.org/about) for their creation of the Rosetta protein structure prediction suite, upon which PyRosetta is built. Julian Rosenberg and J. D. Bagert, former students of the class before PyRosetta, pioneered early drafts of the workshops in 2008 through a Technology Fellowship from the JHU Center for Educational Resources. We thank Richard Shingles of the Center for Educational Resources for assistance in the workshop conception and in the formal assessment over multiple CER grants. Brian Weitzner, Justin Porter, and Liza Lee identified corrections in the first printing. Evan H. Baugh extensively tested the modules for the PyRosetta 2.011 and 2.012 releases, and Jason W. Labonte made corresponding improvements in the text's second edition, respectively. E.H.B. and S.L. developed the PyMOL Mover. J.W.L. contributed Workshop #9. Boonsom Uranukul extensively tested the additions to PyRosetta 3.4, wrote additional sample scripts, and improved the code usability under a Technology Fellowship from the JHU Center for Educational Resources. Alex Mathews, another Technology Fellow, updated the workshops to be current with the newest Talaris2013 scoring function. The National Institutes of Health supported J.J.G., S.C., and S.L. through grant numbers GM-078221 and GM-073151, and the National Science Foundation supported J.J.G. through CAREER award number 0846324. Finally, we thank the wonderful JHU students in all semesters for their help, feedback, patience, and fun times writing code in the lab.

To complete these modules, you will need:

- PyMOL – www.pymol.org
- Python – www.python.org
- IPython – ipython.scipy.org
- PyRosetta – www.pyrosetta.org

All packages are free and available for Mac, Linux, or Windows platforms.

Any opinions, findings, and conclusions or recommendations expressed in this material are those of the authors and do not necessarily reflect the views of the National Science Foundation or the National Institutes of Health.

Workshop #1: PyMOL

PyMOL is a molecular visualization tool. There are many such tools available — both commercially and publicly available:

- PyMOL (www.pymol.org)
- Swiss-PdbViewer (spdbv.vital-it.ch)
- RasMol (www.umass.edu/microbio/rasmol)
- VMD (www.ks.uiuc.edu/Research/vmd)
- MolSim (www.cchem.berkeley.edu/molsim)
- DS Visualizer (accelrys.com/products/discovery-studio/visualization-download.php)
- *etc.*

PyMOL is particularly attractive to us, since it has excellent features for viewing, it is fast and the display quality is superb, it can handle multiple molecules at once, and it is easy to define custom objects such as complexes or sets of atoms. It is also open-source and extensible, so the expert user can create new functions such as colors and measurements related to protein design specifications. The goals of this workshop are to have you become familiar with (1) the basic operation of the software, (2) the tools for analyzing protein structures and for creating high-quality pictures, and (3) the ability to create and save scripts for repeated use.

Suggested Readings

1. Introductory: Chapters 1 and 2 of Brandon and Tooze, *Introduction to Protein Structure*, Garland Publishing (1999).
2. Advanced: J. S. Richardson, "The Anatomy and Taxonomy of Protein Structure," 1981 (updated 2000-2007), available at http://kinemage.biochem.duke.edu/teaching/anatax.

Basic Operations

Download PDB file 1YY8.pdb from www.rcsb.org.

Open PyMOL. (If asked, use the PyMOL + Tcl/Tk mode.) Load 1YY8.pdb. (From the menu bar at the top of the upper window, select File→Open→(select your file).

Use the mouse and mouse buttons to rotate, translate, and zoom the molecule:

On Linux or PC:
left button = rotate, middle button = translate, right button = zoom

On a Mac:
left button = rotate; alt/option+click = translate; right (secondary) click = zoom

The buttons at the top right can set the viewing parameters.
A = Actions, S = Show, H = Hide, L = Label, C = Color.

In the line for 1YY8, select Hide→Everything, then Show→Cartoon, then Color→By Chain→By Chain (e. c), then Label→Chains.

You should see two copies of the antibody fragment, since there were two copies of the antibody in the unit cell of the crystal that was measured to determine the structure. You should be able to see two separate chains in each antibody fragment.

If you click on any atom or residue, the viewer window will display information identifying that atom/residue. Confirm that the four chains are identified as chains A, B, C, and D.

Let's focus in on one fragment. In the command-line window (Depending on your PyMOL version, Windows labels this Tcl/Tk GUI or The PyMOL Molecular Graphics System), type the following commands:

```
select AB, chain A+B
hide all
show cartoon, AB
orient AB
```

Note that in the panel at the top right, you now can operate on the subset AB using the buttons.

You can use the mouse or the `select` command with other protein descriptors (*e.g.*, `name ca+cb+cg+cd`, `symbol o+n`, `resn lys`, `resi 100-150`, `ss h+s+l`, `hydro`, or `hetatm`) to create objects for various subsets of the molecule, and there are a variety of operations you can perform on those subsets.

You can also combine descriptors (`chain A and hydro`) as in the following:

```
select linkerA, chain A and resi 107-112
color red, linkerA
select linkerB, chain B and resi 117-122
color orange, linkerB
```

(Note that there is a distinction in PyMol — as in many scripting/coding languages — between the addition operator + and the Boolean operator `and`. Make sure you understand the difference.)

Type `help select` and `help selections` for full details. (Hit the Esc key to exit the help screen.) Test out the mouse operations and various colorings and display options to get a feel for the general operation of the molecular visualization.

Note that you can use File→Save Session at any time. This will store all your objects, selections, and views.

Structural Analysis

The structure you have downloaded is cetuximab, a therapeutic antibody in development for cancer treatment. Antibodies are composed of two heavy chains and two light chains; this particular construct is known as a Fab fragment and contains one full light chain (chain A) and the N-terminal half of one heavy chain (chain B). At one end of the antibody are six loops known as the "complementarity determining regions" (CDR), which bind a particular antigen. For the following four problems, we will examine the N-terminal domain of chain A (the light chain). To make this easier, type `select L, chain A and resi 1-107`, then from the right panel controls, hide everything but selection L and click Color→Spectrum→Rainbow.

1. Looking down the direction of the first strand, which way does it twist? _____
   Do all strands twist the same direction? _____

2. Next, let's analyze a couple strands in the N-terminal domain. Zoom in on strands 3 and 8, (which should be adjacent and colored cyan and marigold, respectively.) What are the residue <u>number</u> ranges for these two strands? (Click on the strand ends and look in the console window for the residue numbers.)

   strand 3: ____ – ____   strand 8: ____ – ____

   Create a new object for these two strands with `select` and hide the rest of the molecule. Display the atoms of the amino acid and color them by their element (Show→Sticks and Color→ByElement). What color are oxygens? _____ What color are nitrogens? _____ By looking at the side chains, identify the amino-acid sequence (using 1-letter abbreviations) of these two strands:

   strand 3: _____   strand 8: _____

   What is the pattern in these sequences, and why does it occur?

3. Let's analyze some geometry. From the main menu, select Wizard→Measurement. You should see a panel on the right in which you can select distances, angles, dihedrals, and neighbors, and PyMOL will prompt you to select the atoms for measurement. Label→Residues from the right-side panel might also be helpful.

   What is the distance between the N of L73 and the O of F21 (*i.e.*, the hydrogen bonding distance across the β chain)? _____

   Measure all of the backbone hydrogen bonding distances between these two strands. What is the range of distances you observe? _____ – _____

   On residue F21, what is the bond angle around the C_β? _____

On residue F21, measure φ, ψ, and χ₁:

φ: _____ ψ: _____ χ₁: _____

Confirm that these values are within the β-sheet region of the Ramachandran plot.

Type `h_add chain A and resi 73` to place hydrogen atoms on residue 73. (Hydrogens are usually too small to see by crystallography, so PyMOL must calculate the theoretical positions.) What is the H–O–C bond angle for the backbone hydrogen bond between residues L73 and F21? _____

4. Restore the view of L and sketch a "Brandon & Tooze style" topology diagram (*not* a 3-D sketch!) showing the β-sheet strand arrangement for the light chain. Hint: Begin by drawing strand 1 on the bottom left side of the paper pointing up. There are two separate sheets; draw one atop the other; otherwise, it implies a single sheet. Draw the secondary structure first, and then connect the linkers. Do not try to approximate the size of linkers between secondary structure. (p. 62 of Brandon & Tooze has an example of such a two-row topology diagram.)

Comparing Molecules

5. From www.rcsb.org, find a second PDB file of cetuximab, this time bound to its antigen.

   What is the antigen? _____

   Clear your current PyMOL session (All→Actions→Delete everything") and load your new PDB file. Use the cartoon view and color and label by chain to see an overview of the structures. You should see the antibody Fab fragment and the antigen. The antigen also has several post-translational glycosylation modifications.

   Load 1YY8 into the same session. As you did before, create an object for chains A and B and hide chains C and D. (You will now need to specify the molecule: `select unboundFab, 1YY8 and chain A+B`.) Similarly, create an object (Call it boundFab,) for the Fab fragment of the bound complex. (Be careful to specify the correct chain identifiers; they are arbitrary and can vary between PDB files.) Now, superimpose the two structures using `align unboundFab, boundFab`.

The structural match between the two molecules is measured by the root-mean-squared (RMS) distance of the aligned atoms:

$$\text{RMS} = \sqrt{\frac{1}{n}\sum_{i=1}^{n}|\mathbf{x}_i - \mathbf{y}_i|^2}$$

where $\mathbf{x}_i$ and $\mathbf{y}_i$ are the vector coordinates (displacement vectors) of the n atoms in the two structures. The `align` command automatically generates a sequence alignment to pick the right atoms to compare and then solves for and executes the coordinate transformation that yields the minimal RMS deviation between the structures.

6. In the command window, there should be a few lines describing the alignment process. What is the RMS error calculated for this structural alignment (include units)? _____ Over how many atoms? _____

7. Is there much difference between the bound and unbound forms of the antibody? In particular, are there differences in the six complementarity-determining loops at the far end of the N-terminal domains?

High-Quality Visualization and Scripting

Your commands can be saved to a file or read in from a file. Use the File→Log option to record your steps and create a script. You can edit this script using a text editor such as Notepad, WordPad, jEdit, vi, Emacs, or IDLE. You can then read in the script using File→Run or simply with the command `run myfile.pml`. (UNIX commands, such as `pwd`, `ls`, and `cd`, can be typed in PyMOL to assist in locating any scripts you wish to run.) The script will record all of your settings, but not necessarily the transformations you make by reorienting the molecule with the mouse. To record the screen orientation matrices in your script, type `get_view`.

The command `ray` will use a ray-tracing algorithm to compute the lighting on the molecule. (`ray 800,800` will set the image size to 800 × 800 pixels.) Use this before saving an image using File→Save Image to create publication-quality results. Since the natural background color on a piece of paper is white, use the command `bg white` to change the background color (and use less ink!). Other options are under the Display menu; some options that may help include Display→Color Space→CMYK and Display→Depth Cue→On. The menu command Setting→Transparency can also help show depth and occluded molecules, but it is most important to orient the molecule carefully to show features and to hide all but the most relevant parts of the molecule. Finally, you might also try some of the preset settings from the right-side menu under Actions→Preset.

8. For your last task, choose an interesting feature of cetuximab (β-sheet structure, the antibody complementarity-determining regions, a comparison of bound and unbound antibody loops or the CDR H3 loop in detail, the glycosylation on one of the EGFR side chains, *etc*.) and create a beautiful, ray-traced, white-background, publication-quality figure. Color and label protein features and measurements as you feel appropriate. Use the script feature to gather the list of commands that you find optimal for viewing your object. Edit the script to eliminate the non-essential pieces and make the script clean, concise, and comprehensible.

   If you are doing this exercise for a class, submit the figure printed in color, the script that can re-create the figure, and a brief statement of which structural feature your figure is designed to show.

   If you work in a research lab, you are encouraged to create a new figure for a protein relevant to your research.

Workshop #2: PyRosetta

Rosetta is a suite of algorithms for biomolecular structure prediction and design. Rosetta is written in C++ and is available from www.rosettacommons.org. PyRosetta is a toolkit in the programming language Python, which encapsulates the Rosetta functionality by using the compiled C++ libraries. Python is an easy language to learn and also includes modern programming approaches such as objects. It can be used via scripts and also interactively as a command-line program, similar to MATLAB®.

The goals of this first workshop are (1) to have you learn to use PyRosetta both interactively and by writing programs and (2) to have you learn the PyRosetta functions to access and manipulate properties of protein structure.

Basic Elements

You will need a few basic tools to work on PyRosetta.

- You need a text editor to edit scripts. A good editor will "markup" your code in color and make sure your code is indented properly, and it can offer search tools across multiple files and sometimes support for running and debugging your program. One current favorite editor is **jEdit** (www.jedit.org). A popular editor on the Mac is **Aquamacs**, based on the program **Emacs**. **IDLE** is an "integrated development environment" (IDE) that is packaged with Python and includes pop-up function signatures while you are writing code. A text-only (no mouse) program is **vi** or **Vim** (www.vim.org), popular among *nix hackers. jEdit, Emacs, and vi are available for Windows and Linux platforms. There is a built in Mac editor called **TextEdit**, similar to **Notepad** or **WordPad** on a PC. These will not have the color markup and other tools, but they will allow you to edit your files. Choose one of these programs and learn to access it on your computer.
- You need a **command-line interface (CLI)** or **terminal**. On a Windows PC, typing `cmd` under the Start menu will launch a Command Prompt which will support standard DOS commands: `dir`, `cd`, `copy`, `type`, `more`, *etc.* On the Mac, you can find a terminal in the menu on the bottom of the screen or by searching for **xterm**. The Mac terminal will support standard UNIX/Linux shell commands: `ls`, `cd`, `less`, `cp`, `mkdir`, `rm`, `grep`, `awk`, `sed`, `gnuplot`, *etc.* Search the Internet if you are not familiar with Linux shell or DOS commands.
- You can access Python using the command `ipython` or `python ipython.py` from the terminal. We use **IPython** (rather than Python) since it supports tab-completion which will help us find PyRosetta functions. On Windows, your install may include a desktop shortcut for `iPython PyRosetta Shell`: this shortcut will open a terminal and start IPython for you.

Basic Python

Basic Python programming will be useful but is beyond the scope of this workshop. Excellent introductory and reference material on the Python language is available at docs.python.org. A very brief reference is also found in Appendix A.

Basic PyRosetta

1. Open a terminal and start IPython. To load the PyRosetta library, type

    ```
    from rosetta import *
    rosetta.init()
    ```
    or simply `init()`

    The first line loads the Rosetta commands for use in the Python shell, and the second command loads the Rosetta database files. The first line may require a few seconds to load.

2. Many pdb files, like the one you opened in Workshop #1, have extraneous information and often do not conform to file standards. You may have to "clean" your pdb file before loading it into PyRosetta. You can do this through the command line interface (not from within IPython) by using either the `grep` command (UNIX) or the `findstr` command (DOS) to remove all lines that do not begin with `ATOM` in the pdb file. Alternatively, a method from the PyRosetta `toolbox` namespace, `cleanATOM` can be used to create "clean" pdb files:

    ```
    from toolbox import cleanATOM
    cleanATOM("1YY8.pdb")
    ```

    (This method will create a cleaned `1YY8.clean.pdb` file for you.)

    See Appendix C for details on these methods and specific examples of how to clean pdb files.

3. Load a protein from a "clean" pdb file. Use the 1YY8.pdb file of the antibody you looked at in Workshop #1. Put the file in your working directory or change to the directory in which the file is located using `cd` from within IPython. Load the file as follows:

    ```
    pose = pose_from_pdb("1YY8.clean.pdb")
    ```

    This creates a `Pose` object that you can now work with using a variety of methods.

    If you have not already downloaded the pdb file, you can create a pose directly from the protein database if you have a connection to the Internet:

    ```
    from toolbox import pose_from_rcsb
    pose = pose_from_rcsb("1YY8")
    ```

(This method will also create `1YY8.pdb` and `1YY8.clean.pdb` files for you)

4. Examine the protein using a variety of query functions:

    ```
    print pose
    print pose.sequence()
    print "Protein has", pose.total_residue(), "residues."
    print pose.residue(500).name()
    ```

    What type of residue is residue 500? _____

    Note, this is the 500th residue in the pdb file but not necessarily "residue number 500" in the protein. Many pdb files have multiple peptide chains. Sometimes the residue numbering follows a convention from a family of homologous proteins, and often several residues of the N-terminus do not show up in a crystal structure. Find out the chain and pdb residue number of residue 500: _____

    ```
    print pose.pdb_info().chain(500)
    print pose.pdb_info().number(500)
    ```

    Lookup the Rosetta internal number for residue 100 of chain A:

    ```
    print pose.pdb_info().pdb2pose('A', 100)
    ```

    The converse command is:

    ```
    print pose.pdb_info().pose2pdb(25)
    ```

    Get and display the secondary structure of the pose using a `toolbox` method:

    ```
    from toolbox import get_secstruct
    get_secstruct(pose)
    ```

    To demonstrate IPython's *tab-completion* feature, type in `print pose.seq` and hit the tab key. IPython should complete the keyword `sequence` for you. Type `pose.` and hit the tab key, and you should see a list of functions available for `Pose` objects.

    While we are examining the advantages of IPython, try out the built-in help features by typing any one of the following:

    ```
    Pose?
    ?Pose
    help(Pose)
    ```

    Each of these will give a brief description of the Pose class and its purpose. The last form will also give a list of function signatures for all the available functions within the class. These methods of accessing help should work on many of the PyRosetta objects.

Protein Geometry

5. Find the φ, ψ, and χ₁ dihedral angles of residue 5:

   ```
   print pose.phi(5)
   print pose.psi(5)
   print pose.chi(1, 5)
   ```

6. Find the N–C$_\alpha$ and C$_\alpha$–C bond lengths of residue 5. There are at least a couple ways to do this.

   First, store the unique atom identifier codes in variables:

   ```
   R5N  = AtomID(1, 5)
   R5CA = AtomID(2, 5)
   R5C  = AtomID(3, 5)
   ```

   (This works because the atoms are listed in a consistent order in a pdb file.) Then, use these identifier codes to lookup bond lengths in the conformation object:

   ```
   print pose.conformation().bond_length(R5N, R5CA)
   print pose.conformation().bond_length(R5CA, R5C)
   ```

   For the second method, access the Cartesian coordinates and use the `Vector` class to find the norm of the displacement vector between the two atoms:

   ```
   N_xyz = pose.residue(5).xyz("N")
   CA_xyz = pose.residue(5).xyz("CA")
   N_CA_vector = CA_xyz - N_xyz
   print N_CA_vector.norm
   ```

   These bond lengths are actual, experimental bond lengths from the crystal structure. When Rosetta creates proteins *de novo*, it uses ideal values, similar to those from Engh & Huber (1991). Let's check how the actual bond lengths compare to Rosetta's ideal values. Find the Rosetta database directory on your computer (*e.g.*, `/usr/local/PyRosetta/rosetta_database`). Enter the subdirectory `chemical/residue_type_sets/fa_standard/residue_types` and, with your text editor, load the `param` file appropriate for residue 5. The `ICOOR_INTERNAL` lines give the internal coordinates for an ideal conformation, including the torsion angle, bond angle, and bond length needed to build each subsequent atom in the group.

7. Can you identify the N–C$_\alpha$ and C$_\alpha$–C bond lengths? How do they compare? Bonus: how do they compare to Engh & Huber's numbers? If they differ, why?

8. Find the N–C$_\alpha$–C bond angle in radians:

   ```
   print pose.conformation().bond_angle(R5N, R5CA, R5C)
   ```

   What is this angle in degrees? _____

   Again, compare with the Rosetta database ideal value. What is the hybridization of the C$_\alpha$ atom? _____ What is the standard bond angle for such a hybridization? _____

Be aware that not all bond lengths and angles are accessible through the conformation object. The conformation object only contains a minimal subset of bond lengths and angles used in generating Cartesian coordinates. The vector objects provide a general way to measure angles, distances, and torsions between arbitrary atoms.

9. How could you also find the N–C$_\alpha$–C bond angle using the vector dot product function, `v3 = v1.dot(v2)`? (Recall from vector calculus that the angle between any two displacement vectors $\overrightarrow{BA}$ and $\overrightarrow{BC}$ is $\arccos \frac{\overrightarrow{BA} \cdot \overrightarrow{BC}}{|\overrightarrow{BA}||\overrightarrow{BC}|}$.)

Manipulating Protein Geometry

10. We can also alter the geometry of the protein. Perform each of the following manipulations, and give the coordinates of the N atom of residue 6 afterward.

    ```
    pose.set_phi(5, -60)
    pose.set_psi(5, -43)
    pose.set_chi(1, 5, 180)

    pose.conformation().set_bond_length(R5N, R5CA, 1.5)
    pose.conformation().set_bond_angle(R5N, R5CA, R5C,
                                      110./180.*3.14159)
    ```

    New coordinates of N atom of residue 6: (_____, _____, _____)

Remember that only some bond lengths and angles are available through the conformation object. Note that even though dihedral angles are set in degrees, the bond angle is set in radians! (To make the conversion between degrees and radians easier, you may wish to import Python's `math` module. See Appendix A for more information.)

Visualization and the PyMOL Mover

What if we wish to view the changes to geometry that we have made? We can "dump" the information in a pose object into a new pdb file with the method `pose.dump_pdb("filename.pdb")` and then open this pdb file in our favorite visualization software. However, constantly dumping output and loading new files into a visualizer can be cumbersome; thus, the visualization process was streamlined with the `PyMOL_Mover`, which provides a means for observing structural changes almost instantaneously in PyMOL as they are made in PyRosetta.

First, we must open PyMOL and run a script that will cause PyMOL to listen for instructions from the PyMOL mover. (Certain Windows installations will use a shortcut that automatically does this for you, or you can create a `.pymolrc` file in your home directory in Linux or Mac that runs the code for you):

```
cd <your_PyRosetta_install_path>
run PyMOLPyRosettaServer.py
```

Then, with PyRosetta, we must instantiate a `PyMOL_Mover` and then apply it to a pose any time we make a change:

```
pymol = PyMOL_Mover()
pymol.apply(pose)
```

11. Make some changes to the dihedral angles of a pose and apply the PyMOL mover to watch the effect of the new angles on the structure.

For more advanced PyMOL mover options, visit www.pyrosetta.org/pymol_mover-tutorial, or see Appendix A.

Programming

12. You can write programs in Python to accomplish more complicated tasks. Using your text editor, open a new file with extension `.py` (*e.g.*, `rama.py`). You can write your entire program here and then run it either from the command line by typing

```
[linux]> python rama.py
```

or from inside a Python shell by typing

```
In [1]: run rama.py
```

Here is a sample program:

```
from rosetta import *
init()
p = pose_from_pdb("1ABC.pdb")

for i in range(1, p.total_residue() + 1):
    print i, "phi =", p.phi(i), "psi =", p.psi(i)
```

Note that we must always first import the Rosetta modules with the `import` command and initialize Rosetta with the `init()` command before loading a pose. (Appendix A contains a list of common Python commands and syntax.) Test that you can write and run a simple program from a file.

Programming Exercises

Submit your script file and your output.

1. *Torsion angle.* Use the vector objects to write a script to calculate torsion angles between four arbitrary atoms. This will require knowledge of vector calculus. Hint: You will need to calculate the normal vectors of the two planes of the dihedral angle.

2. *Ideal helix.* Write a program to create a 20-residue ideal helix by setting the φ and ψ angles to the typical values for an α-helix. To start, use `pose = pose_from_sequence("AAA", "fa_standard")` to create a new pose, except use 20 "A"s in the argument to create a 20-residue poly-alanine. Output your structure using `pose.dump_pdb("helix.pdb")`.

   View your new file in PyMOL to check your work. How can you be sure your structure is a proper α-helix? List three distinct structural characteristics that you can check.

3. *Ideal strand.* Write a program to create a 20-residue ideal β-strand by setting the φ and ψ angles to values in the middle of the β region of the Ramachandran plot.

   View your new file in PyMOL to check your work. How can you be sure your structure is a proper β-strand? List three distinct structural characteristics that you can check.

4. *Secondary structure propensities.* Write a program to calculate the propensity of each residue type to appear in a helix. Loop through all residues in a protein, and count each alanine that is in a helix, sheet, or loop according to some φ/ψ-based criteria. The propensity can then be calculated as $P_{\alpha\,Ala} = \frac{N_{\alpha\,Ala}}{N_{\alpha\,total}}$, $P_{\beta\,Ala} = \frac{N_{\beta\,Ala}}{N_{\beta\,total}}$, and $P_{L\,Ala} = \frac{N_{L\,Ala}}{N_{L\,total}}$, where $N_{\alpha\,Ala}$, $N_{\beta\,Ala}$, and $N_{L\,Ala}$, are the counts of alanine residues in helices, sheets, and loops, respectively, and $N_{\alpha\,total}$, $N_{\beta\,total}$, and $N_{L\,total}$, are the counts of *all* residues in helices, sheets, and loops, respectively. (Note that terminal residues have different names in Rosetta than internal ones; *e.g.*, an N-terminal `ALA` has the name `ALA_p:NtermProteinFull`.)

Bonus level 1: Find propensities for all 20 amino acid types. This will be easier if you use a data structure (list, array, dictionary, map) to store the counts of the 20 types. Do the residues with the highest helical propensity match that given by Brandon & Tooze?

Bonus level 2: To get better statistics, collect your data by looping over a set of 10 pdb files. Better yet, use a set of files such as the PDBSelect set of representative chains (http://bioinfo.tg.fh-giessen.de/pdbselect; this may require considerable download and compute time.)

5. *Idealize a protein.* Write a program that sets all bond lengths and angles to their Engh & Huber ideal values. Test your program using a structure from the pdb. What happens to the resulting protein? Why?

6. *Cleaning pdb files.* Coordinate files in the Protein Data Bank are quite diverse, and many pdb files have variations in their format to accommodate peculiarities such as post-translationally modified residues or disordered regions where coordinates could not be determined for certain atoms. In addition, some pdb files simply do not conform to the file standards. When the pdb file departs from the standards, it is necessary to clean-up the pdb file before loading it into Rosetta. (See Appendix C.) For this exercise, examine PDB ID 1D4I (HIV-1 protease in complex with an inhibitor).

What non-standard amino acid is present, and what is this amino acid? (Hint: examine the pdb file header or the web page summary.)

Write a script that converts the non-standard amino acid to its unmodified form. (Hint: use UNIX commands `grep`, `awk`, or `sed` along with pipes. Note: It is also possible to directly use a modified amino acid by creating a parameter file to define that residue, but that is a more advanced topic!)

answer: The following UNIX shell command will change ABA to ALA and change the HETATM keys to ATOM (enter as a single-line command!):

```
awk '{if ($1 == "HETATM" && $4 == "ABA")
                         {gsub("HETATM","ATOM");
                          gsub("ABA","ALA")};
                          print}' 1D41.pdb | grep
                          ^ATOM > 1D41.clean.pdb
```

References

1. R. A. Engh & R. Huber, "Accurate bond and angle parameters for X-ray protein structure refinement," *Acta. Cryst. A* **47**, 392–400 (1991).
2. J. Parsons *et al.*, "Practical conversion from torsion space to Cartesian space for *in silico* protein synthesis," *J. Comp. Chem.* **26**, 1063–1068 (2005).
3. Python help available at http://docs.python.org.

Workshop #3: PyRosetta Scoring

Scoring Poses

A basic function of Rosetta is calculating the energy or *score* of a biomolecule. Rosetta has a standard energy function for all-atom calculations as well as several scoring functions for low-resolution protein representations. In addition, you can tailor an energy function by including scoring terms of your choice with custom weights.

For these exercises, use the protein Ras (PDB ID 6Q21) and load it into a pose called `ras`. Be sure to clean the PDB file and use only one chain.

1. To score a protein, you must first define a scoring function:

    ```
    scorefxn = get_fa_scorefxn()
    ```

    The method `get_fa_scorefxn` tells Rosetta to load the default full-atom energy terms. To see these terms, you can print the score function:

    ```
    print scorefxn
    ```

    What terms are in the score function, and what are their relative weights?

    (Appendix A includes a list of Rosetta scoring function terms, which you may wish to reference.)

2. Set up your own custom score function that includes just van der Waals, solvation, and hydrogen bonding terms, all with weights of 1.0. Use the following to start:

    ```
    scorefxn2 = ScoreFunction()
    scorefxn2.set_weight(fa_atr, 1.0)
    scorefxn2.set_weight(fa_rep, 1.0)
    ```

    (The first line declares the object `scorefxn2` as part of the `ScoreFunction` class.) Enter the other weights and then confirm that the weights are set correctly.

3. Evaluate the energy of Ras with the standard score function:

    ```
    print scorefxn(ras)
    ```

    What is the total energy of Ras? _____

    Break the energy down into its individual pieces with the `show` method:

    ```
    scorefxn.show(ras)
    ```

    Which are the three most dominant contributions, and what are their values? Is this what you would have expected? Why?

4. Unweighted, individual component energies of each residue in a structure are stored in the `Pose` object and can be accessed by its `energies()` object. For example, to break the energy down into each residue's contribution use:

    ```
    print ras.energies().show(n)
    ```

    where `n` is the residue number.

    What are the total van der Waals, solvation, and hydrogen bonding contributions of residue 24? (Note that the *backbone* hydrogen bonding terms for each residue are *not* available from the `Energies` object.)

5. The van der Waals and solvation terms are atom–atom pairwise energies calculated from a pre-tabulated lookup table, dependent on the distance between the two atoms and their types. You can access the lookup table (`etable`) directly to check these energy calculations on an atom-by-atom basis:

   ```
   r1 = ras.residue(24)
   r2 = ras.residue(20)
   a1 = r1.atom("N")
   a2 = r2.atom("O")
   etable_atom_pair_energies(a1, a2, scorefxn)
   ```

   (Note that the `etable_atom_pair_energies()` function requires `Atom` objects, not the `AtomID` objects we saw in Workshop #2.)

   What are the attractive, repulsive, and solvation components between the nitrogen of residue 24 and the oxygen of residue 20?

   How does Rosetta separate the "attractive" and "repulsive" van der Waals components? (Hint: it is *not* the r^{-6} and r^{-12} parts of the Lennard–Jones equation.)

6. The hydrogen-bonding score component requires identification of acceptor hybridization state and calculation of geometric parameters including distance, acceptor bond angle, proton bond angle, and a torsion angle. The hydrogen-bonding energies are stored in an `HBondSet` object. You can access the list of hydrogen bonds by creating an `HBondSet` object, filling the set from the pose (after making sure the pose has had its `Energies` object updated based on neighboring residues within the pose), and then using the `HBondSet.show()` command:

   ```
   hbond_set = hbonds.HBondSet()
   ras.update_residue_neighbors()
   hbonds.fill_hbond_set(ras, False, hbond_set)
   hbond_set.show(ras)
   ```

   (Note that some functions and classes, such as `HBondSet`, must be referenced with their proper "namespace", such as the namespace `hbonds` here.)

   The above steps have been combined in the PyRosetta `toolbox` method `get_hbonds()`, so that we can also simply type:

   ```
   from toolbox import get_hbonds
   hbond_set = get_hbonds(ras)
   hbond_set.show(ras)
   ```

An individual residue's hydrogen bonds can be looked up from the set using its residue number:

```
hbond_set.show(ras, 24)
```

How many hydrogen bonds does residue 24 make? _____

Using the parameters from the show function, sketch a picture of this hydrogen bond. Label all possible atoms, distances and angles.

7. Analyze the energy between residues Y102 and Q408 in cetuximab (1YY9). (You'll need to load that structure into a new `Pose` object.

   a. As explained in Workshop #2, internally, a `Pose` object has a list of residues, numbered starting from 1. To find the residue numbers of Y102 of chain D and Q408 of chain A, use the residue chain identifier and the PDB residue number to convert to the pose numbering:

   ```
   print pose.pdb_info().pdb2pose('A', 102)
   ```

   D:Y102: _____   A:Q408: _____

   b. Score the pose and determine the Van der Waals energies and solvation energy between these two residues. Use the following commands to isolate contributions from particular pairs of residues, where `rsd1` and `rsd2` are the two residue objects of interest (not the residue number — use `pose.residue(res_num)` to access the objects):

   ```
   emap = EMapVector()
   scorefxn.eval_ci_2b(rsd1, rsd2, pose, emap)
   print emap[fa_atr]
   print emap[fa_rep]
   print emap[fa_sol]
   ```

   c. How do Rosetta's calculations compare to what you might calculate by hand? (See references.)

d. Create a new PDB file containing coordinates for just the two residues Y102 and Q408. Repeat the above calculations. Which energies change? Why?

Energies and the PyMOL Mover

The `PyMOL_Mover` class contains a method for sending score function information to PyMOL, which will then color the structure based on relative residue energies.

8. Instantiate a `PyMOL_Mover` and then use `pymol.send_energy(pose)` to send the coloring command to PyMOL after scoring `ras` with `scorefxn`.
   What color is residue Pro34? _____
   What color is residue Ala66? _____
   Which residue has a lower energy? _____

9. `pymol.send_energy(pose, fa_atr)` will have PyMOL color only by the attractive van der Waals energy component. What color is residue 34 if colored by *solvation* energy? _____

If its `update_energy` option is true, the PyMOL mover will update the colors by energy every time the mover is applied to the pose:

```
pymol.update_energy(True)
pymol.energy_type(fa_atr)
pymol.apply(pose)
```

You can also have PyMOL label each Cα with the value of its residue's energy:

```
pymol.label_energy(pose, fa_rep)
```

Finally, if you have scored the pose first, you can have PyMOL display all of the calculated hydrogen bonds for the structure:

```
pymol.send_hbonds(pose)
```

Programming Exercises

1. *Interface energy*. Write a program that can calculate the binding energy of EGFR to cetuximab. You will need to make separate PDB files for the antigen, antibody, and complex. In PyMOL, select one of these peptides, then use File→Save Molecule.

   Use the following formula for binding energy:

   $$E_{binding} = E_{complex} - E_{antibody} - E_{antigen}$$

Submit your script along with its output, which should include values for the total binding energy, along with the *weighted* contributions to the binding energy from Van der Waals, hydrogen bonding, and solvation. What does your result suggest about these two proteins *in vitro*? What are some inaccuracies in the way you've calculated the binding energy?

2. *Statistical energy functions.* Write a program to output a file of data of the C–N bond lengths observed in a set of ten high-resolution X-ray protein structures. (One source of curated structures is the WHATIF sets at http://swift.cmbi.kun.nl/swift/whatif/select.)

   a. Import the file into a spreadsheet, and plot the data as a histogram of probability versus bond length and also as a statistical energy ($E = -kT \ln P$, where P is probability and kT is set to 0.6 kcal/mol.) versus bond length. Try a bin size of 0.01 Å.

   b. Look up the CHARMm parameters for this bond stretch, and plot a curve over the statistical energy vs. bond length figure to show the CHARMm model for this motion.

   c. Do the statistics match what would be produced by a harmonic oscillator under the CHARMm potential? Specifically, are the average bond length and the CHARMm spring constant K correct? If not, what should it be? You may need to fit a parabola to your data to find the average bond length and the spring constant K.

3. Write a program to loop over all pairs of atoms in two residues to confirm that the sum of the individual atom–atom pair energies (calculated using the `Etable` lookup) is the same as the total residue-residue pair energy.

References

1. E. Neria, S. Fischer & M. Karplus, "Simulation of activation free energies in molecular systems," *J. Chem. Phys.* **105**, 1902-1921 (1996).
2. T. Kortemme, A. V. Morozov & D. Baker, "An orientation-dependent hydrogen bonding potential improves prediction of specificity and structure for proteins and protein-protein complexes," *J. Mol. Biol.* **326**, 1239-1259 (2003).
3. D. Eisenberg & A. D. McLachlan, "Solvation energy in protein folding and binding," *Nature* **319**, 199-203 (1986).
4. T. Lazaridis & M. Karplus, "Effective energy function for proteins in solution," *Proteins* **35**, 133-152 (1999).

Workshop #4: PyRosetta Folding

In this workshop you will write your own Monte Carlo protein folding algorithm from scratch, and we will explore a couple of the tricks used by Simons *et al.* (1997, 1999) to speed up the folding search.

Suggested Readings

1. K. T. Simons *et al.*, "Assembly of Protein Structures from Fragments," *J. Mol. Biol.* **268**, 209-225 (1997).
2. K. T. Simons *et al.*, "Improved recognition of protein structures," *Proteins* **34**, 82-95 (1999).
3. Chapter 4 (Monte Carlo methods) of M. P. Allen & D. J. Tildesley, *Computer Simulation of Liquids*, Oxford University Press, 1989.

A Simple *de Novo* Folding Algorithm

1. First, we would like to create a simple folding algorithm. Begin with a new pose, and then create an all-atom starting structure with 10 alanines using:

   ```
   pose = pose_from_sequence('A'*10)
   ```

   Use the `PyMOL_Mover` to echo this structure to PyMOL:

   ```
   pmm = PyMOL_Mover()
   pmm.apply(pose)
   ```

   You should see ideal bond lengths and angles, although the set of φ and ψ angles will not be useful.

   Now, write a program that implements a Monte Carlo algorithm to optimize the protein conformation. In the main program, create a loop with 100 iterations. Each iteration should call a subroutine to make a random trial move, score the protein, and then accept or reject the new conformation based on the Metropolis criterion, for which the probability of accepting a move is $P = e^{-\Delta E/kT}$, when $\Delta E \geq 0$, and $P = 1$, when $\Delta E < 0$. Use $kT = 1$ Rosetta Energy Unit.

   For the random trial move, write a *subroutine* to choose one residue at random using `random.randint()` and then randomly perturb either the φ or ψ angles by a random number chosen from a Gaussian distribution. Use the Python built-in function `random.gauss()` from the `random` library with a mean of the current angle and a standard deviation of 25°. After changing the torsion angle, use `pmm.apply(pose)` to update the structure in PyMOL.

For the energy function, use the standard full-atom scoring approach with *only* the van der Waals and hydrogen bonding terms. With this scoring function, what kind of structures do you expect to be most stable?

At each iteration of the search, output the current pose energy and the lowest energy ever observed. The final output of this program should be the lowest energy conformation that is achieved at any point during the simulation. Be sure to use `low_pose.assign(pose)` rather than `low_pose = pose`, since the latter will only copy a pointer to the original pose.

Output the last pose and the lowest-scoring pose observed and view them in PyMOL. Plot the energy and lowest-energy observed vs. cycle number. What are the energies of the initial, last, and lowest-scoring pose? Is your program working? Has it converged to a good solution?

2. Using the program you wrote for Workshop #2, force the A_{10} sequence into an α-helix. Does this structure have a lower score than that produced by your algorithm? What does this mean about your sampling or discrimination?

3. Since your program is a stochastic search algorithm, it may not produce an ideal structure consistently, so try running the simulation multiple times or with a different number of cycles (if necessary). Using a *kT* of 1, your program may need to make up to 500,000 iterations.

Low-Resolution (Centroid) Scoring

Following the treatment of Simons *et al.* (1999), Rosetta can score a protein conformation using a low-resolution representation. This will make the energy calculation faster.

4. Load a protein with which you are familiar (*e.g.*, Ras or cetuximab). Calculate the full-atom energy and note the coordinates of residue 5 using `print pose.residue(5)`.

$E:$ _____ (_____ , _____ , _____)

5. Convert the pose to the centroid form by using a `SwitchResidueTypeSetMover` object and the `apply` method:

   ```
   switch = SwitchResidueTypeSetMover("centroid")
   switch.apply(pose)
   print pose.residue(5)
   ```

   How many atoms are now in residue 5? _____ How is this different than before?

6. Score the new, centroid-based pose using the standard score function `"score3"`. What is the new total score? _____ What scoring terms are included in `"score3"`? Do these match Simons?

7. Convert the pose back to all-atom form by using another switch mover (`SwitchResidueTypeSetMover("fa_standard")`). Confirm that you have all the atoms back. Are the atoms in the same position as before? _____

8. Adjust your folding algorithm to use centroid residue types. Use only `vdw` and `hbond_sr_bb` energy score components. How much faster does your program run?

Not counting the `PyMol_Mover`, which is a special case, `SwitchResidueTypeSetMover` is the first example we have seen of a `Mover` class in PyRosetta. Every `Mover` object in PyRosetta has been designed to apply specific and complex changes (or "moves") to a pose. Every mover must be "constructed" and have any options set before being applied to a pose with the `apply()` method. `SwitchResidueTypeSetMover` has a relatively simple construction with only the single option `"centroid"`. (Some movers, as we shall see, require no options and are programmed to operate with default values.)

Protein Fragments

9. Create a 3-mer fragment file from the Robetta server (http://robetta.bakerlab.org/fragmentsubmit.jsp) for a given test sequence of at least 26 residues of your choice. Before submitting your job, select **Exclude homologs**. When your job is complete, download the file `aatestA03_05.200_v1_3`. This file contains 3-mer fragments for the test sequence we are trying to fold. You should see sets of three-lines describing each fragment. For the first fragment, which PDB file does it come from? _____ Is this fragment helical, sheet, in a loop, or a combination? _____ What are the φ, ψ, and ω angles of the middle residue of the first fragment window?

   φ: _____ ψ: _____ ω: _____

10. How many 3-residue windows are there in your 20-residue peptide? ____ How many fragments does the data file have per window? _____

11. Create a new subroutine in your folding code for an alternate random move based upon a "fragment insertion". A fragment insertion is the replacement of the torsion angles for a set of consecutive residues with new torsion angles pulled at random from a fragment library file. Prior to calling the subroutine, load the set of fragments from the fragment file:

    ```
    fragset = ConstantLengthFragSet(3)
    fragset.read_fragment_file("aatestA03_05.200_v1_3")
    ```

    Next, we will construct another `Mover` object — this time a `FragmentMover` — using the above fragment set and a `MoveMap` object as options. A move map specifies which degrees of freedom are allowed to change in the pose when the mover is applied (in this case, all backbone torsion angles):

    ```
    movemap = MoveMap()
    movemap.set_bb(True)
    mover_3mer = ClassicFragmentMover(fragset, movemap)
    ```

    (Note that when a `MoveMap` is constructed, all degrees of freedom are set to `False` initially. If you still have a PyMOL_Mover instantiated, you can quickly visualize which degrees of freedom will be allowed by sending your move map to PyMOL with `pmm.send_movemap(pose, movemap)`.)

    Each time this mover is applied, it will select a random 3-mer window and insert only the backbone torsion angles from a random matching fragment in the fragment set:

    ```
    mover_3mer.apply(pose)
    ```

When you change your random move to a fragment insertion, how much faster is your folding code? Does it converge to a protein-like conformation more quickly?

Programming Exercises

1. Fold a 10-mer poly-alanine using 100 independent trajectories, using any variant of the folding algorithm that you like. (A trajectory is a path through the conformation space traveled during the calculation. The end result of each independent trajectory is called a "decoy". Given enough sampling, the lowest energy decoy may correspond to the global minimum.) Create a Ramachandran plot using the lowest-scoring conformations (decoys) from all 100 independent trajectories. Repeat this for a 10-mer poly-glycine. How do the plots differ? Compare with the plots in Richardson's article.

2. Test your folding program's ability to predict a real fold from scratch. Choose a small protein to keep the computation time down, such as Hox-B1 homeobox protein (1B72) or RecA (2REB). How many iterations and how many independent trajectories do you need to run to find a good structure?

3. Modify your folding program to include a simulated annealing temperature schedule, decaying exponentially from $kT = 100$ to $kT = 0.1$ over the course of the search. Again, fold a test protein. Does this approach work better?

4. Modify your folding program to remove the Metropolis criterion and instead accept trial moves *only* when the energy decreases. Plot energy vs. iteration and examine the final output structures from multiple runs. How is the convergence and performance affected? Why?

Thought Questions

1. [Introductory] What are the limitations of these types of folding algorithms?

2. [Advanced] How might you design an intermediate-resolution representation of side chains that has more detail than the centroid approach yet is faster than the full-atom approach? Which types of residues would most benefit from this type of representation?

Workshop #5: PyRosetta Refinement

One of the most basic operations in protein structure and design algorithms is manipulation of the protein conformation. In Rosetta, these manipulations are organized into movers. A `Mover` object simply changes the conformation of a given pose. It can be simple, like a single φ or ψ angle change, or complex, like an entire refinement protocol.

Suggested Reading

1. P. Bradley, K. M. S. Misura & D. Baker, "Toward high-resolution de novo structure prediction for small proteins," *Science* **309**, 1868-1871 (2005), including Supplementary Material.
2. Z. Li & H. A. Scheraga, "Monte Carlo-minimization approach to the multiple-minima problem in protein folding," *Proc. Natl. Acad. Sci. USA* **84**, 6611-6615 (1987).

Introduction

In the last workshop, you encountered the `ClassicFragmentMover`, which inserts a short sequence of backbone torsion angles, and the `SwitchResidueTypeSetMover`, which doesn't actually change the conformation of the pose but instead swaps out the residue types used.

In this workshop, we will introduce a variety of other movers, particularly those used in high-resolution refinement (*e.g.*, in Bradley's 2005 paper).

Before you start, load a test protein and make a copy of the pose so we can compare later:

```
start = pose_from_pdb("test.pdb")
test = Pose()
test.assign(start)
```

For convenient viewing in PyMOL, set the names of both poses:

```
start.pdb_info().name("start")
test.pdb_info().name("test")

pmm = PyMOL_Mover()
pmm.apply(start)
pmm.apply(test)
```

We also want to activate the `keep_history` setting so that PyMOL will keep separate frames for each conformation as we modify it (more on this shortly):

```
pmm.keep_history(True)
```

Small and Shear Moves

The simplest move types are small moves, which perturb φ or ψ of a random residue by a random small angle, and shear moves, which perturb φ of a random residue by a small angle and ψ of the same residue by the same small angle of opposite sign.

For convenience, the small and shear movers can do multiple rounds of perturbation. They also check that the new φ/ψ combinations are within an allowable region of the Ramachandran plot by using a Metropolis acceptance criterion based on the `rama` score component change. (The `rama` score is a statistical score from Simons *et al.* 1999, parametrized by bins of φ/ψ space.) Because they use the Metropolis criterion, we must also supply *kT*. Finally, like most movers, these require a `MoveMap` object to specify which degrees of freedom are fixed and which are free to change. Thus, we can create our movers like this:

```
kT = 1.0
n_moves = 1
movemap = MoveMap()
movemap.set_bb(True)
small_mover = SmallMover(movemap, kT, n_moves)
shear_mover = ShearMover(movemap, kT, n_moves)
```

We can also adjust the maximum magnitude of the perturbations as follows:

```
small_mover.angle_max("H", 25)
small_mover.angle_max("E", 25)
small_mover.angle_max("L", 25)
```

Here, `"H"`, `"E"`, and `"L"` refer to helical, sheet, and loop residues — as they did in the fragment library file — and the magnitude is in degrees. We will set all the maximum angles to 25° to make the changes easy to visualize. (The default values in Rosetta are 0°, 5°, and 6°, respectively.)

1. Test your mover by applying it to your pose:

    ```
    small_mover.apply(test)
    ```

    Confirm that the change has occurred by comparing the `start` and `test` poses in PyMOL. Second, try the PyMOL animation controls on the bottom right corner of the Viewer window. There should be a play button (▶) as well as frame-forward, rewind, *etc*. Play the movie to watch PyMOL shuffle your pose move back and forth.

    Can you identify which torsion angles changed? By how much? If it is hard to view on the screen, it may help to use your old programs to compare torsion angles or coordinates.

2. *Comparing small and shear movers.* Reset the test pose by re-assigning it the conformation from `start`, and create a second test pose (`test2`) in the same manner. Reset the existing `MoveMap` object to *only* allow the backbone angles of residue 50 to move. (Hint: Set all residues to `False`, then set just residues 50 and 51 to `True` using `movemap.set_bb(50, True)` and `movemap.set_bb(51, True)`.) Note that the `SmallMover` contains a pointer to your `MoveMap`, and so it will automatically know you have made these changes and use the modified `MoveMap` in future moves.

   Make one small move on one of your test poses and one shear move on the other test pose. Output both poses to PyMOL using the `PyMOL_Mover`. Be sure to set the name of each pose so they are distinguishable in PyMOL. Show only backbone atoms and view as lines or sticks. Identify the torsion angle changes that occurred. What was the magnitude of the change in the sheared pose? How does the displacement of residue 60 compare between the small- and shear-perturbed poses?

Minimization Moves

The `MinMover` carries out a gradient-based minimization to find the nearest local minimum in the energy function, such as that used in one step of the Monte-Carlo-plus-Minimization algorithm of Li & Scheraga.

```
min_mover = MinMover()
```

3. The minimization mover needs at least a `MoveMap` and a `ScoreFunction`. You can also specify different minimization algorithms and a tolerance. (See Appendix A). For now, set up a new `MoveMap` that is flexible from residues 40 to 60, inclusive, using:

   ```
   mm4060 = MoveMap()
   mm4060.set_bb_true_range(40, 60)
   ```

   Create a standard, full-atom `ScoreFunction`, and then attach these objects to the default `MinMover` with the following methods:

   ```
   min_mover.movemap(mm4060)
   min_mover.score_function(scorefxn)
   ```

   Finally, attach an "observer". The observer is configured to execute a `PyMOL_Mover.apply()` every time a change is observed in the pose coordinates. The `True` is a flag to ensure that PyMOL keeps a history of the moves.

   ```
   observer = PyMOL_Observer(test2, True)
   ```

4. Apply the `MinMover` to your sheared pose. Observe the output in PyMOL. (This may take a couple minutes — the Observer can slow down the minimization significantly.) How much motion do you see, relative to the original shear move? How many coordinate updates does the `MinMover` try? How does the magnitude of the motion change as the minimization continues? At the end, how far has the C_α atom of residue 60 moved?

Monte Carlo Object

PyRosetta has several object classes for convenience for building more complex algorithms. One example is the `MonteCarlo` object. This object performs all the bookkeeping you need for creating a Monte Carlo search. That is, it can decide whether to accept or reject a trial conformation, and it keeps track of the lowest-energy conformation and other statistics about the search. Having the Monte Carlo operations packaged together is convenient, especially if we want multiple Monte Carlo loops to nest within each other or to operate on different parts of the protein. To create the object, you need an initial pose, a score function, and a temperature factor:

```
mc = MonteCarlo(pose, scorefxn, kT)
```

After the pose is modified by a mover, we tell the `MonteCarlo` object to automatically accept or reject the new conformation and update a set of internal counters by calling:

```
mc.boltzmann(pose)
```

5. Test out a `MonteCarlo` object. Before doing so, you may need to adjust your small and shear moves to smaller maximum angles (3–5°) so they are more likely to be accepted. Apply several small or shear moves, output the score using `print scorefxn(test)` then call `mc.boltzmann(test)`. A response of `True` indicates the move is accepted, and `False` indicates that the move is rejected. If the move is rejected, the pose is automatically reverted for you to its last accepted state. Manually iterate a few times between moves and calls to `mc.boltzmann()`. Do enough cycles to observe at least two `True` and two `False` outputs. Do the acceptances match what you expect given the scores you obtain? _____ After doing a few cycles, use `mc.show_scores()` to find the score of the last accepted state and the lowest energy state. What energies do you find? Is the last accepted energy equal to the lowest energy?

6. See what information is stored in the Monte Carlo object using:

   ```
   mc.show_scores()
   mc.show_counters()
   mc.show_state()
   ```

   What information do you get from each of these?

Trial Mover

A `TrialMover` combines a mover with a `MonteCarlo` object. Each time a `TrialMover` is called, it performs a trial move *and* tests that move's acceptance with the `MonteCarlo` object. You can create a `TrialMover` from any other type of `Mover`. You might imagine that, as we start nesting these together, we can build some complex algorithms!

   ```
   trial_mover = TrialMover(small_mover, mc)
   trial_mover.apply(pose)
   ```

7. Apply the `TrialMover` above ten times. Using `trial_mover.num_accepts()` and `trial_mover.acceptance_rate()`, what do you find?

8. The `TrialMover` also communicates information to the `MonteCarlo` object about the type of moves being tried. Create a second `TrialMover` using a `ShearMover` and the same `MonteCarlo` object, and apply this second `TrialMover` ten times. Look at the `MonteCarlo` object state (`mc.show_state()`). What are the acceptance rates of each mover? Which mover is accepted most often, and which has the largest energy drop per trial? What are the average energy drops?

Sequence and Repeat Movers

A `SequenceMover` is another combination `Mover` and applies several movers in succession. It is useful for building up complex routines and is constructed as follows.

```
seq_mover = SequenceMover()
seq_mover.add_mover(small_mover)
seq_mover.add_mover(shear_mover)
seq_mover.add_mover(min_mover)
```

The above example mover will apply first the small, then the shear, and finally the minimization movers.

9. Create a `TrialMover` using the sequence mover above, and apply it five times to the pose. How is the sequence mover shown by `mc.show_state()`?

A `RepeatMover` will apply its input `Mover` n times each time it is applied:

```
repeat_mover = RepeatMover(trial_mover, n)
```

10. Use these tools to build up your own *ab initio* protocol. Create trial movers for 9-mer and 3-mer fragment insertions. First, create repeat movers for each and then create the trial movers using the same `MonteCarlo` object for each. Create a `SequenceMover` to do the 9-mer trials and then the 3-mer trials, and iterate the sequence 10 times. Write out a flowchart of your algorithm here:

11. *Hierarchical search.* Construct a `TrialMover` that tries to insert a 9-mer fragment and then refines the protein with 100 alternating small and shear trials before the next 9-mer fragment trial. The interesting part is this: you will use one `MonteCarlo` object for the small and shear trials, inside the whole 9-mer combination mover. But use a separate `MonteCarlo` object for the 9-mer trials. In this way, if a 9-mer fragment insertion is evaluated after the optimization by small and shear moves and is rejected, the pose goes all the way back to before the 9-mer fragment insertion.

Refinement Protocol

The entire standard Rosetta refinement protocol, similar to that presented in Bradley, Misura, & Baker 2005, is available as a `Mover`. Note that the protocol can require ~40 minutes for a 100-residue protein.

```
relax = ClassicRelax()
relax.set_scorefxn(scorefxn)
relax.apply(pose)
```

Programming Exercises

1. Use the `Mover` constructs to create a complex folding algorithm. Create a simple program to do the following:

    a. Five small moves
    b. Minimize
    c. Five shear moves
    d. Minimize
    e. Monte Carlo Metropolis criterion
    f. Repeat a–e 100 times
    g. Repeat a–f five times, each time decreasing the magnitude of the small and shear moves from 25° to 5° in 5° increments.

   Sketch a flowchart, and submit both the flowchart and your code.

2. Ab initio *folding algorithm.* Based on the Monte Carlo energy optimization algorithm from Workshop #4, write a complete program that will fold a protein. A suggested algorithm involves preliminary low-resolution modifications by fragment insertion (first 9-mers, then 3-mers), followed by high-resolution refinement using small, shear, and minimization movers. Output both your low-resolution intermediate structure and the final refined, high-resolution decoy.

   Test your code by attempting to fold domain 2 of the RecA protein (the last 60 amino acid residues of PDB ID 2REB). How do your results compare with the crystal structure? (Consider both your low-resolution and high-resolution results.) If your lowest-energy conformation is different than the native structure, explain why this is so in terms of the limitations of the computational approach.

Bonus: After using the `PyMOL_Mover` or `PyMOL_Observer` to record the trajectory, export the frames and tie them together to create an animation. Search the Internet for "PyMOL animation" for additional tools and tips. Animated GIF files are probably the best quality; MPEG and QuickTime formats are also popular and widely compatible and uploadable to YouTube.

3. *AraC N-terminal arm.* The AraC transcription factor is believed to be activated by the conformational change that occurs in the N-terminus when arabinose binds. Let's test whether PyRosetta can capture this change. Specifically, we will start with the arabinose-bound form and see if PyRosetta can refold it to the apo form.

    Download the arabinose-bound form of the AraC transcription factor. Edit the PDB file so that it contains only the arabinose-binding domain, and also remove any non-protein atoms (especially the arabinose). Set up a move map to include only the 15 N-terminal residues. Perform an *ab initio* search to find the lowest conformation state. How does it compare to the apo crystal form?

Thought Questions

1. With $kT = 1$, what is the change in propensity of the `rama` score component that has a 50% chance of being accepted as a small move?

2. How would you test whether an algorithm is effective? That is, what kind of measures can you use? What can you vary within an algorithm to make it more effective?

Workshop #6: Packing & Design

Rosetta uses a Monte Carlo optimization routine to pack side chains using a library of conformations, or rotamers. This operation can be used for side-chain packing for operations like refinement or for designing optimal sequences. This workshop will examine both capabilities.

Suggested readings

1. J. Desmet *et al.*, "The dead-end elimination theorem and its use in protein side-chain positioning," *Nature* **356**, 539-543 (1992).
2. B. Kuhlman & D. Baker, "Native protein structures are close to optimal for their structures," *PNAS* 97, 10383, 2000.

Side Chain Conformations, the Rotamer Library, and Dunbrack Energies

Begin by loading cetuximab from PDB 1YY8.

1. What are the φ, ψ, and χ angles of residue K49?

   φ: _____ ψ: _____ χ_1: _____ χ_2: _____ χ_3: _____ χ_4: _____

2. Open `asp.bbdep.rotamers.lib` by unpacking `asp.bbdep.rotamers.gz` from the directory `rosetta_database/rotamer/ExtendedOpt1-5`. Find the φ/ψ bin for lysine at residue 49 and find the nearest rotamer. What are the χ angles and standard deviations of this rotamer? What is its probability?

   χ_1: _____ ± _____ χ_2: _____ ± _____ χ_3: _____ ± _____ χ_4: _____ ± _____

   $P =$ _____

3. Score your pose with the standard full-atom score function. What is the energy of K49? Note the Dunbrack energy component (`fa_dun`), which represents the side-chain conformational probability. Does it match what you found in the table? (You will need to convert between probability and energy; use *kT* = 1.) If not, why not?

4. Use `pose.set_chi(i, res_num, chi)` to set the side chain of residue 49 to the all-anti conformation. (Here, `i` is the χ index, and `chi` is the new torsion angle in degrees.) Re-score the pose and note the Dunbrack energy. Does it match the probability in the table? _____ Is this conformation valid for cetuximab (*i.e.*, is the total score of this residue acceptable)? _____

Monte Carlo Side-Chain Packing

Side-chain packing can be done in a Monte Carlo search routine that iteratively swaps rotamers of a random residue and tests each move using the Metropolis criterion. Rosetta has such a routine pre-packaged as a `Mover` that carries out a simulated annealing search each time it is applied. The specific scope of the packing is specified in a `PackerTask` object, which is similar to a `MoveMap` in that it specifies degrees of freedom. We can specify via commands or from an input file our settings for a `PackerTask`. Create a PackerTask as follows. This will set the task to allow packing only of residue 49:

```
task_pack = standard_packer_task(pose)
task_pack.restrict_to_repacking()
task_pack.temporarily_fix_everything()
task_pack.temporarily_set_pack_residue(49, True)
```

The default task allows *any* amino acid residue to be swapped in for another; that is, it would simulate a protein variant as a result of mutation. This would be useful for protein design but not for side-chain packing. `restrict_to_repacking()` only allows rotamers from the current residue at that position to be used. We can confirm our settings using

```
print task_pack
```

We now can construct a `PackRotamersMover`:

```
pack_mover = PackRotamersMover(scorefxn, task_pack)
```

5. Apply the `PackMover` above to your pose with `pack_mover.apply(pose)`. Now what are the χ angles of K49? Which rotamer is this? What is the Dunbrack energy?

6. What is the new total energy of K49? _____ Why did Rosetta pick this rotamer? Answer this in terms of the components of the score function and in terms of the residues with which K49 interacts.

Packing for Refinement

Side-chain packing can be used when converting a pose from centroid to full-atom mode, and it is used extensively in full-atom refinement calculations. Let's examine how packing improves scores.

Use your code from Workshop #5 to create a centroid-representation model for RecA protein domain 2. Save that centroid "decoy" so that we can compare several basic refinement steps.

7. Load the centroid decoy and convert it to full-atom representation using the `SwitchResidueTypeSetMover`. Save this starting configuration for future use. Score the pose with the standard score function. Why is the score so high?

8. Create a default `PackRotamersMover` with a `PackerTask` that allows all residues to vary χ angles. Create a test pose from your start pose and pack the side chains. What is the new pose score? _____

9. Reset the test pose to the start configuration. Create a `MinMover` using the Davidson-Fletcher-Powell minimization scheme by applying the method `min_type("dfpmin")` to your mover. Create a `MoveMap` that allows χ angles but *not* φ/ψ/ω angles to vary. Apply the `MinMover` and rescore the pose. How does this energy compare?

10. Again, reset the test pose. Apply the packer and then minimize on the χ angles. Now what is the final score? _____

For fun, you might examine the individual residue energies to find the residues most responsible for the score changes. Typically, a small number of residues may make clashes that can be resolved using the χ angle minimization, which allows off-rotamer side-chain conformations.

Design

Design calculations can be accomplished simply by packing side chains with a rotamer set that includes all amino acid types. That is, when the Monte Carlo routine swaps rotamers, it could replace the existing side chain with another conformation of the same residue or some conformation of a different residue type. Trial mutations are accepted or rejected with the Metropolis criterion, and the standard full-atom energy function is supplemented by a reference energy term, `ref`, which represents the relative energies of each residue type in an unfolded peptide.

Design operations are easiest to specify through a data file called a "resfile." You can create a resfile for a given pdb file or pose using the following `toolbox` methods:

```
from toolbox import generate_resfile_from_pdb
generate_resfile_from_pdb("1YY8.pdb", "1YY8.resfile")
from toolbox import generate_resfile_from_pose
generate_resfile_from_pose(pose, "1YY8.resfile")
```

Inside the resfile you will see a list of all residues and their chain with `NATRO` next to that, indicating that the position is set to use the <u>na</u>tive <u>ro</u>tamer. `NATRO` can be changed to any of the following with a text editor:

`NATRO`	use native amino acid and native rotamer (does not repack)
`NATAA`	use native amino acid but allow repacking to other rotamers
`PIKAA ILV`	use only the following amino acids and allow repacking between them
`ALLAA`	use all amino acids and all repacking

Edit the resfile to force residue 49 to be glutamic acid (`49 A PIKAA E`) and save the file as `1YY8-K49E.resfile`. Create a new task for design from the resfile:

```
task_design = TaskFactory.create_packer_task(pose)
parse_resfile(pose, task_design, "1YY8-K49E.resfile")
```

11. Score the original conformation from the pdb for reference. Create a new `PackResiduesMover` with the design task and use it to mutate residue 49 to glutamic acid. What is the predicted ΔG of the mutation? _____ Is this a stabilizing mutation? _____

12. Note the residue reference energy term (`ref`) in the scoring function. What is this value before and after you mutated the residue? What does this energy represent?

13. Create a new `PackerTask` and `PackRotamersMover` using a new resfile that allows residue 49 to be designed freely (`49 A ALLAA`), and apply the mover. What residue does Rosetta choose? _____ Why?

14. Create your own resfile that will restrict residue 49 to only negatively charged residues using the resfile line `49 A PIKAA DE` and re-apply the design mover. Now what residue is chosen? _____ What is the new residue energy, and why (physically) is it less favorable than the last design?

15. Let's try to make this design more favorable. Select several surrounding residues for design, and set them also to enable mutations to any residue. Call the design mover again. Now what do you find?

Workshop #6: Packing & Design | 41

It should be noted that PyRosetta includes a handy `toolbox` method `mutate_residue()` that will change a specified residue in a given pose into another. However, the rotamer of this new residue will not be optimized. For example:

```
from toolbox import mutate_residue
mutate_residue(pose, 49, 'E')
```

Programming Exercises

1. *Refinement and discrimination.* Download the "single misfold" decoy set from the Decoys 'R Us repository at dd.compbio.washington.edu/ddownload.cgi?misfold. (Documentation for this project is at dd.compbio.washington.edu.) This repository has a single "correct" and "incorrect" predicted structure for several proteins. For this exercise, analyze pdbs 2CI2 and 2CRO; each has two "incorrect" structures offered. (Technical note: These decoys have an empty occupancy field in the PDB ATOM records; a value of 1 needs to be added before Rosetta will load these structures.)

   Write a program that will calculate and output the score for each decoy (i) as is from the PDB file, (ii) after packing only, (iii) after minimization only, and (iv) after packing and minimizing. For each of the four cases, compare the scores of the "correct" structure with those of the "incorrect" structure. Which schemes successfully discriminate the correct structures?

2. Write a refinement protocol that will iterate between side-chain packing, small and shear moves, and minimization. Where is the best place to position the Monte Carlo acceptance test? Test your protocol by making 10 independently-refined structures for the correct and incorrect decoys of 2CRO from the Decoys 'R Us single misfold set. Is this protocol able to discriminate the correct decoy? Submit your code.

3. HIV-1 protease is a major drug target for antiretroviral therapies. Protease inhibitors are designed from substrate peptide mimics. We will attempt to take a natural substrate peptide of HIV-1 protease and design it for improved binding — potentially to serve as a good template for drug design. Use PDB file 1KJG for the following analysis.

   a. Turn on side-chain packing for the protease active site (residues 8, 23, 25, 29, 30, 32, 45, 47, 50, 53, 82, and 84 of both chains A and B) and for the peptide (residues 2–9 on chain P; all of these numbers follow the PDB numbering).

   b. Repack the above side chains and then energy minimize those same side chains with the backbone fixed. Generate 10 decoys and record the energies for each decoy. This will represent the reference state: the wild-type peptide bound to the protease.

   c. For residue 2 of the peptide (chain P), allow repacking to any of the 20 amino acid residues, while leaving the packing and side-chain minimization the same as in step b. Generate 10 decoys and record the energies. These will represent single mutants at that residue position.

d. Repeat step c for each of the other 8 residues in the substrate peptide.

e. Take the lowest energy for each mutation position and compare it to the lowest energy for the wild type. Do single mutants at any of these positions improve the energy over the wild type? Which ones? By how much? Which energy components are mostly responsible?

f. Which peptide residue positions are easiest to improve? Which positions are the hardest?

g. Are there any other trends? Hydrophobic *vs*. polar, bulky residues *vs*. small residues, *etc*.?

h. Altman *et al.* (*Proteins* 2008) found, using their own computational design algorithm, that the most favorable sequences were a triple mutant E3D/T4I/V6L, a single mutant T4V, and a single mutant E3Q. How do their results compare with yours?

i. Natural substrates are often sub-optimal binders. Why would this be advantageous?

4. *Effect of backbone conformation on design*. HIV-1 protease is promiscuous, meaning it can cleave a wide range of peptides beyond the ten natural substrates of the virus. Let's examine the preferences of the enzyme through Rosetta design calculations.

   a. Download HIV-1 protease in complex with CA-P2 peptide (1F7A). Select the eight peptide residues for unrestricted design and let Rosetta redesign the substrate sequence. What is the new sequence and how does it compare to the original? What percent of the original sequence was optimal for its structure?

   b. Download HIV-1 protease in complex with RT-RH peptide (1KJG). (Note that the enzyme is the same here, but it is crystallized with a different substrate.) Again, design the eight substrate residues with Rosetta. What percent of this substrate sequence is optimal for this crystal structure? ____%

   c. How do the designed sequences of (a) and (b) compare? Why should they be the same? Why would they not be the same? What are the implications for the field of computational protein design?

5. Write a program which iterates between design of all residues of a protein and refinement via small, shear, and minimization moves.

Thought Question

1. What is the thermodynamic meaning of the `ref` energy term, and what does it correspond to *physically*?

2. During evolution, the genome sequence may mutate to cause protein sequence changes. Alternately, one could consider the difference in evolutionary propensities for each residue type. How could you derive reference energies from sequence data, and what would that mean?

3. How do Kuhlman & Baker fit the reference energies in their 2000 *PNAS* paper?

References

1. S. C. Lovell *et al.*, "The penultimate rotamer library," *Proteins* **40**, 389-408 (2000).
2. R. L. Dunbrack & F. E. Cohen, "Bayesian statistical analysis of protein side-chain rotamer preferences," *Protein Sci.* **6**, 1661-1681 (1997)

Workshop #7: Docking

Protein–protein docking is the prediction of a complex structure starting from its monomer components. The search space can be extremely large, so large amounts of computational resources are typically required. In this workshop, we will explore several of the techniques briefly; keep in mind that for real applications, many more decoys will need to be tested.

Suggested Readings

1. J. J. Gray *et al.*, "Protein-protein docking with simultaneous optimization of rigid-body displacement and side-chain conformations," *J. Mol. Biol.* **331**, 281-299 (2003).
2. S. Chaudhury & J. J. Gray, "Conformer selection and induced fit in flexible backbone protein-protein docking using computational and NMR ensembles," *J. Mol. Biol.* **381**, 1068-1087 (2008).

Fast Fourier Transform Based Docking via ZDOCK

There are several servers available based on fast Fourier transforms (FFTs). These servers are able to quickly carryout a global, grid-based matching searches.

1. Go to the ZDOCK server (http://zdock.bu.edu) and upload trypsin (2PTN) and its inhibitor (1BA7 chain B) for docking. If completing this workshop for a class, do this in groups in order to not overload the server. When the jobs have finished (typically under an hour), download the output file. You will have to also download a script for creating complexes from the output file. Use the script to generate the top five models. Are these models similar or diverse? _____ How so?

2. Are any of the models similar to the crystal structure of the bound complex (1AVW)? _____

(Other servers include SmoothDock (http://structure.pitt.edu/servers/smoothdock), ClusPro (http://cluspro.bu.edu), Haddock (http://haddock.chem.uu.nl), and GRAMM-X (http://vakser.bioinformatics.ku.edu/resources/gramm/grammx). Any of these provide global docking services to create models that might be useful for refinement by RosettaDock.)

Docking Moves in Rosetta

For the following exercises, download and clean the complex of colicin D and ImmD (1V74). Store three poses — a full-atom starting pose and centroid and full-atom "working" poses.

The fundamental docking move is a rigid-body transformation consisting of a translation and rotation. Any rigid body move also needs to know which part moves and which part is fixed. In Rosetta, this division is known as a "jump" and the set of protein segments and jumps are stored in an object attached to a pose called a "fold tree."

```
print pose.fold_tree()
```

In the fold tree printout, each three number sequence following the word EDGE is the beginning and ending residue number, then a code. The codes are -1 for stretches of protein and any positive integer for a jump, which represents the jump number.

3. View the fold tree of your full-atom pose. How many jumps are there in your pose? ___

By default, there is a jump between the N-terminus of chain A and the N-terminus of chain B, but we can change this using the exposed method setup_foldtree().

```
setup_foldtree(pose, "A_B", Vector1([1]))
print pose.fold_tree()
```

The argument "A_B" tells Rosetta to make chain A the "rigid" chain and allow chain B to move. If there were more chains in the pdb structure, supplying "AB_C" would hold chains A and B rigid together as a single unit and allow chain C to move. (The third argument Vector1([1]) is required, it creates a Rosetta vector object — indexed from 1 — with one element that identifies the first jump in the fold tree for docking use.)

4. Set up a new fold tree for docking using the command above and output the new fold tree. What has changed?

You can see the type of information in the jump by printing it from the pose:

```
jump_num = 1
print pose.jump(jump_num).get_rotation()
print pose.jump(jump_num).get_translation()
```

5. Write out the rotation matrix and the translation vector defined by the jump.

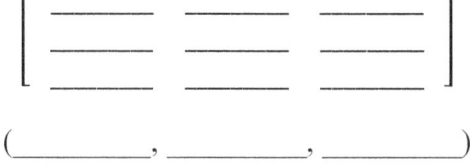

The two basic manipulations are translations and rotations. For translation, the change in *x*, *y*, and *z* coordinates are needed as well as the jump number. A rotation requires a center and an axis about which to rotate. The rigid-body displacement can be altered directly with the RigidBodyTransMover for translations or the RigidBodySpinMover for rotations.

However, for structure prediction calculations, we have a mover that is preconfigured to make random movements varying around set magnitudes (in this case, a mean of 8° rotation and 3 Å translation) located in the `rosetta.protocols.rigid` namespace, (which we will rename with an alias `rigid_moves` for our convenience) :

```
import rosetta.protocols.rigid as rigid_moves
pert_mover = rigid_moves.RigidBodyPerturbMover(jump_num,
                                               8, 3)
```

6. Apply the `RigidBodyPerturbMover` to a pose and use a `PyMOL_Mover` to confirm that the motions are what you expect. What are the new rotation matrix and translation vector in the jump? How many ångströms did the downstream protein move?

Global perturbations are useful for making completely randomized starting structures. The following mover will rotate a protein about its geometric center. The final orientation is equally distributed over the "globe".

```
randomize1 = rigid_moves.RigidBodyRandomizeMover(pose,
                            jump_num,
                            rigid_moves.partner_upstream)
randomize2 = rigid_moves.RigidBodyRandomizeMover(pose,
                            jump_num,
                            rigid_moves.partner_downstream)
```

(`partner_upstream` and `partner_downstream` are predefined terms within the `rosetta.protocols.rigid` namespace, which in our case refer to chains A and B, respectively.)

7. Apply both movers to the starting structure, and view the structure in PyMOL. (You might view it along with the original pose.) Does the new conformation look like a candidate docked structure yet? _____

Since proteins are not spherical, sometimes the random orientation creates severe clashes between the docking partners, and other times it places the partners so they are no longer touching. The `FaDockingSlideIntoContact` mover will translate the downstream protein along the line of protein centers until the proteins are in contact.

```
slide = DockingSlideIntoContact(jump_num)   # for centroid
                                              mode
slide = FaDockingSlideIntoContact(jump_num)  # for full-
                                              atom mode
slide.apply(pose)
```

The `MinMover`, which we have previously used to change torsion angles to find the nearest minimum in the score function, can also operate on the jump translation and rotation. It suffices to set the jump variable as moveable in the `MoveMap`:

```
movemap = MoveMap()
movemap.set_jump(jump_num, True)

min_mover = MinMover()
min_mover.movemap(movemap)
min_mover.score_function(scorefxn)   # use any scorefxn
scorefxn(pose)
min_mover.apply(pose)
```

8. Apply the above `MinMover`. How much does the score change? _____ What are the new rotation matrix and translation vector in the jump? How many Ångstroms did the downstream protein move?

Low-Resolution Docking via RosettaDock

RosettaDock can also perform global docking runs, but it can require significant time. Typically, 10^5 to 10^6 decoys are needed in a global run. For this workshop, we will create a much smaller number and learn the tools needed to handle large runs.

Docking is available as a mover that completely encompasses the protocol. To use the mover, you will need a starting pose with both chains and a jump defined. The structure must be in low-resolution (centroid) mode, and you will need the low-resolution docking score function:

```
scorefxn_low = create_score_function("interchain_cen")
```

Randomize your centroid version of the complex. Then, create low-resolution docking structures as follows:

```
dock_lowres = DockingLowRes(scorefxn_low, jump_num)
dock_lowres.apply(pose)
```

9. You can compare structures by calculating the root-mean-squared deviation of all the C_α atoms, using the function `CA_rmsd(pose1, pose2)`. In docking, a more useful measure is the ligand RMSD, which is the deviation of the backbone C_α atoms of the ligand after superposition of the receptor protein backbones. You can calculate ligand RMSD with `calc_Lrmsd(pose1, pose2, Vector1([1]))`. Using both measures, how far did your pose move from the low-resolution search?

10. Examine the created decoy in PyMOL. Does it look like a reasonable structure for a protein-protein complex? _____ Explain.

Job Distributor

For exhaustive searches with Rosetta (docking, refinement, or folding), it is necessary to create a large number of candidate structures, termed "decoys". This is often accomplished by spreading out the work over a large number of computers. Additionally, each decoy created needs to be individually labeled. The object that is responsible for managing the output is called a `JobDistributor`. Here, we will use a simple job distributor to create multiple structures. The following constructor sets the job distributor to create 10 decoys, with filenames like `output_1.pdb`, `output_2.pdb`, *etc*. The pdb files will also include scores according to the `ScoreFunction` provided.

```
jd = PyJobDistributor("output", 10, scorefxn_low)
```

It is also useful to compare each decoy to the native structure (if it is known; otherwise any reference structure can be used). The job distributor will do the RMSD calculation and final scoring upon output. To set the native pose:

```
native_pose = pose_from_pdb("your_favorite_protein.pdb")
jd.native_pose = native_pose
```

11. Create a randomized starting pose, working pose, fold tree, score function, job distributor, and low-resolution docking mover. Now, run the low-resolution docking protocol to create a structure, and output a decoy:

    ```
    pose.assign(starting_pose)
    dock_lowres.apply(pose)
    jd.output_decoy(pose)
    ```

    Do this twice and confirm that you have two output files.

Whenever the `output_decoy()` method is called, the `current_num` variable of the `JobDistributor` is incremented, and it also outputs an updated score file: `output.fasc`. We can finish the set of 10 decoys by using the `JobDistributor` to set up a loop:

```
while not jd.job_complete:
    pose.assign(starting_pose)
    dock_lowres.apply(pose)
    jd.output_decoy(pose)
```

Note the `jd.job_complete` Boolean variable that indicates whether all 10 decoys have been created.

12. Run the loop to create 10 structures. The score file, `output.fasc` summarizes the energies and RMSDs of all structures created. Examine that file. What is the lowest score? _____ What is the lowest energy? _____

13. Reset the `JobDistributor` to create 100 decoys (or more or less, as the speed of your processor allows) by reconstructing it. Rerun the loop above to make 100 decoys. Use your score file to plot score versus RMSD. (Two easy ways to do this are to import the score file into Excel or to use the Linux command gnuplot.) Do you see an energy funnel? _____

High-Resolution Docking

The high-resolution stage of RosettaDock is also available as a mover. This mover encompasses random rigid-body moves, side-chain packing, and gradient-based minimization in the rigid-body coordinates. High-resolution docking needs an all-atom score function. The optimized docking weights are available as a patch to the standard all-atom energy function.

```
scorefxn_high = create_score_function_ws_patch(
                                    "talaris2013",
                                    "docking")
dock_hires = DockMCMProtocol()
dock_hires.set_scorefxn(scorefxn_high)
dock_hires.set_partners("A_B")
```

Note that unlike for `DockingLowRes`, we must supply the docking partners with `"A_B"` instead of `jump_num`.

A high-resolution decoy needs side chains. One way to place the side chains is to call the `PackMover`, which will generate a conformation from rotamers. A second way is to copy the side chains from the original monomer structures. This is often helpful for docking calculations since the monomer crystal structures have good side chain positions.

```
recover_sidechains = ReturnSidechainMover(starting_pose)
recover_sidechains.apply(pose)
```

14. Load one of your low-resolution decoys, add the side chains from the starting pose, and refine the decoy using high-resolution docking. How far did the structure move during refinement? _____ How much did the score improve? _____

15. Starting from your lowest-scoring low-resolution decoy, create three high-resolution decoys. (You might use the `JobDistributor`.) Do the same starting from the native structure.

    a. How do the refined-native scores compare to the refined-decoy scores?

    b. What is the RMSD of the refined native? _____ Why is it not zero?

    c. How much variation do you see in the refined native scores? In the refined decoy scores? Is the difference between the refined natives and the refined decoys significant?

Docking Funnel

Using a job distributor and `DockMCMProtocol`, create 10 decoys starting with a `RigidBodyRandomizeMover` perturbation of `partner_downstream`, 10 decoys starting from different local random perturbations (8°, 3 Å), 10 decoys starting from low-resolution decoys, and 10 starting from the native structure. Plot all of these points on a funnel plot. How is the sampling from each method? Does the scoring function discriminate good complexes?

Programming Exercises

1. Output a structure with a 10 Å translation and another with a 30° rotation (both starting from the same starting structure), and load them into PyMOL to confirm the motions are what you expect.

2. *Diffusion*. Make a series of random rigid body perturbations and record the RMSD after each. Plot RMSD versus the number of moves. Does this process emulate diffusion? If it did, how would you know? (Hint: there is a way to plot these data to make them linear.)

3. Starting from a low-resolution docking decoy, refine the structure in three separate ways:

    a. side-chain packing
    b. gradient-based minimization in the rigid-body coordinates
    c. gradient-based minimization in the torsional coordinates
    d. the docking high-resolution protocol

    For each, note the change in RMSD and the change in score. Which operations move the protein the most? Which make the most difference in the score?

4. Using the `MonteCarlo` object, the `RigidBodyMover`, `PackRotamers`, and the `MinMover`, create your own high-resolution docking protocol. Bonus: Can you tune it to beat the standard protocol? "Beating" the standard protocol could mean achieving lower energies, running in faster time, and/or being able to better predict complexes.

Workshop #8: Loop Modeling

Loop modeling is an important step in building homology models, designing enzymes, or docking with flexible loops.

Suggested Readings

1. A. A. Canutescu & R. L. Dunbrack, "Cyclic coordinate descent: A robotics algorithm for protein loop closure," *Protein Sci.* **12**, 963-972 (2003).
2. C. Wang, P. Bradley & D. Baker, "Protein-protein docking with backbone flexibility," *J. Mol. Biol.* **373**, 503-519 (2007).
3. D. J. Mandell, E. A. Coutsias & T. Kortemme, "Sub-angstrom accuracy in protein loop reconstruction by robotics-inspired conformational sampling," *Nature Meth.* **6**, 551-552 (2009).

Fold Tree

Because we typically want to isolate the conformational changes to the loop region, we need a framework to hold the rest of the protein together. This is accomplished — as was the case with protein-protein docking — with a *fold tree*, which is a graph that dictates the propagation of conformational changes throughout the `Pose`.

For the following exercises, you can download the loop modeling package from the Gray lab website (http://graylab.jhu.edu/pyrosetta/downloads/pyrosetta_scripts/loop_modeling.zip). There you will find `test_in.pdb` and a 3mer fragment file, `test_in.frag3`.

Load the `test_in.pdb` structure (116 residues). We want to operate on the first loop, residues 15–24. For the fold tree, we place the jump anchors two residues outside of the loop range, *i.e.*, residues 13–26. In loop modeling, the jump points are set at *i*-2 and *j*+2, where *i* and *j* are the beginning and end residues of the loop, respectively. The relevant fold tree looks like this:

```
                    ┌─────────┐
         ─────────  │         ──────────────────────────────
         1     13 19│20  26                              116
```

That is, we want a cut between residues 19 and 20, to allow motions in the loop that do not propagate through the rest of the protein. To tie the pieces together, we use a jump between residues 13 and 26. These residues will stay connected to each other.

To make such a tree in PyRosetta, first we create a `FoldTree` object:

```
ft = FoldTree()
```

Then we add the "edges" and the jump. Both edges and jumps are entered using the `FoldTree`'s `add_edge(start, end, code)` command, with peptide edges coded with a "-1" and jumps enumerated with the positive integers. (The first jump is coded "1", the second "2", *etc*.) The first edge is from residues 1 to 13:

```
ft.add_edge(1, 13, -1)
```

Then the second is from 13 to 19. An edge must always start from a residue that has already been defined in another edge, thus we use 13 here and not 14. (The one exception is the first edge, which starts from the graph's "root").

```
ft.add_edge(13, 19, -1)
```

Next, the jump, which is specified with the integer code 1, tells Rosetta that this is a rigid-body connection, not a peptide edge:

```
ft.add_edge(13, 26, 1)
```

Finally, we add the last two edges, both starting from 26, which is the residue that has been previously defined in the tree:

```
ft.add_edge(26, 20, -1)
ft.add_edge(26, 116, -1)
```

Print the fold tree and check that this tree is valid:

```
print ft
ft.check_fold_tree()
```

The latter command will return `False` if there are any invalid connections, disconnected parts, or undefined residues.

Finally, we attach this fold tree to the pose, overwriting its default fold tree:

```
pose.fold_tree(ft)
```

1. Test out your fold tree. Do `pose.set_phi(res, 180)` for `res` values of 10, 13, 16, 23, 26 and 29. View each in PyMOL with the PyMOL mover or viewer. (It may be helpful to color the structure by its foldtree using the `PyMOL_Mover.send_foldtree()` method.) What do you observe in these structures?

2. Sketch a fold tree that you could use for modeling a loop from residues 78–83. Remember that a loop from residues *i* to *j* uses a fold tree with a jump from residues *i*-2 to *j*+2.

3. What edges would you use to generate the above fold tree?

4. Visualize the foldtree that you just sketched above in PyMOL using `PyMOL_Mover.view_foldtree_diagram(pose, ft)`. Does your sketch match what is drawn on the screen? _____

To save some time and help avoid mistakes, there are a couple functions that will assist in the creation of fold trees:

5. Try each of the following and print the fold tree. What does each of the following do?

    ```
    ft.clear()
    ft.simple_tree(116)
    ft.new_jump(76, 85, 80)
    ```

6. Use these `FoldTree` methods to check your answer to question 3.

7. Use the above commands to make a fold tree to model both loops (15–24 and 78–83) simultaneously.

Cyclic Coordination Descent (CCD) Loop Closure

Canutescu & Dunbrack's CCD routine is implemented as a `Mover`. It first requires that the loop is defined using the `Loop` class. You will also need to create a `MoveMap` with the loop residues marked as flexible (both backbone and side-chain torsion angles). Like many specialty objects in Rosetta, the CCD mover is located in its own namespace and is not loaded in by default when the rest of Rosetta is imported into Python. You can either import the module or refer to its namespace when calling it. Here we show the module import.

```
from rosetta.protocols.loops.loop_closure.ccd import *

loop1 = Loop(15, 24, 19)
ccd = CcdLoopClosureMover(loop1, movemap)
```

8. Open the loop using `set_phi`, and run the CCD Mover. Does it close the loop? _____ Is the bond across the cut point protein-like? _____

Note also that if you have a loop defined in a Loop object, you can set your fold tree with the command:

```
set_single_loop_fold_tree(pose, loop)
```

Multiple Loops

Multiple loops can be stored in a `Loops` object. We can create a `loop2` object for the 78–83 loop and then create a `loops` object:

```
loops = Loops()
loops.add_loop(loop1)
loops.add_loop(loop2)
```

To use CCD on all loops, we have to iterate over each one.

Loop Building

The `MoveMap` and the `FoldTree` work together. By using a `MoveMap`, you can ensure that a `Mover` will only operate inside the loop region.

9. At this point, you should be able to write your own loop protocol that will build the loop at low-resolution using fragments. Some tips:

    - Create a `MoveMap` that will allow motions only in the two loop regions defined in our `MoveMap` above.
    - Create a `ClassicFragmentMover` using your `MoveMap` and the 3-residue fragment file provided, `test_in.frag3`.

- Use the `cen_std` score function, but add the `chainbreak` score component with a weight of 1.
- Do 100 fragment insertions.
- After each fragment insertion, close the loop with CCD, then use a `MonteCarlo` object to accept or reject the combination move.
- Bonus: use `SequenceMover` and `TrialMover` to tighten up your code.
- Further bonus: use the `JobDistributor` to allow your program to make multiple structures.

Loop RMSD is typically measured in a fixed reference frame of the whole protein, and it can be computed on Cα atoms or all backbone atoms. PyRosetta has a built-in function for calculating deviation of all the loops, and its output can be added as additional info in the `JobDistributor`:

```
lrms = loop_rmsd(pose, reference_pose, loops, True)
jd.additional_decoy_info = " LRMSD: " + str(lrms)
```

(The fourth argument in `loop_rmsd()` tells whether or not the RMSD is calculated for C_α atoms only.)

10. If you first perturb the loop residues by setting all the residues to extended conformations ($\varphi=\psi=180°$), can your code close the two loops and find reasonable conformations? _____ What is the loop RMS? _____

High-Resolution Loop Protocol

In high-resolution, loop optimization needs smaller perturbations such as that from small and shear moves. The classic Rosetta loop refinement protocol is available as a `Mover` and is located in the rosetta.protocols.loops.loop_mover.refine namespace:

```
from rosetta.protocols.loops.loop_mover.refine import *

loop_refine = LoopMover_Refine_CCD(loops)
```

The mover uses its own default, high-resolution score function, and it will generate its own `MoveMap` based on the definition of the loops.

11. Apply this mover to a few of your low-resolution loop models after switching them to full atom versions. How far does refinement move the loops? _____ Do the loops remain closed? _____

Kinematic Closure (KIC) Protocols

The kinematic loop closure methods of Mandell *et al.* are also available in several prepared movers and protocols.

A single closure of a loop using the direct mathematical solution using polynomial resultants can be performed by applying the `KinematicMover`. Rather than using a `Loop` object, this mover is set up by specifying the three "pivot" residues. The kinematic mover determines new torsion angles for the pivot residues by solving the closure equations.

```
from rosetta.protocols.loops.loop_closure.kinematic_closure
                                         import *
kic_mover = KinematicMover()
kic_mover.set_pivots(16, 20, 24)
kic_mover.apply(pose)
```

Like the `CcdLoopClosureMover`, the `KinematicMover` can be applied after various perturbations that alter or open the loop.

The full loop prediction protocol of Mandell *et al.* has been implemented as separate movers for the low- and high-resolution stages. The `LoopMover_Perturb_KIC` operates on a centroid representation pose, and it is designed to predict a loop *de novo*. The `LoopMover_Refine_KIC` operates in the full-atom representation, and it is designed to refine a loop making small perturbations from a starting conformation such as one output by the `LoopMover_Perturb_KIC` mover. Here is an example of the usage of both:

```
from rosetta.protocols.loops.loop_mover.perturb import *
from rosetta.protocols.loops.loop_mover.refine import *

sw = SwitchResidueTypeSetMover("centroid")
sw.apply(pose)
kic_perturb = LoopMover_Perturb_KIC(loops)
kic_perturb.apply(pose)

sw = SwitchResidueTypeSetMover("fa_standard")
sw.apply(pose)

kic_refine = LoopMover_Refine_KIC(loops)
kic_refine.apply(pose)
```

Because these are full prediction protocols, they will require some time to perform on your computer. Each will result in a single decoy structure; note that Mandell *et al.* generated 1000 decoy structures for blind predictions of 12-residue loops.

12. Time permitting, repeat exercises 8–10, replacing the CCD mover with the appropriate kinematic movers.

Simultaneous Loop Modeling and Docking

Antibodies have two chains, light (L) and heavy (H), which can move relative to each other. They also have a long, hypervariable H3 loop, typically residues 95–102. Antibodies are common protein drugs, and they are often created by exploiting the immune system of a mouse. There is a need for high-quality homology models of antibodies.

13. Sketch a fold tree that you could use to model an antibody with a flexible H3 loop and H and L chains that can move relative to each other.

14. Write a low-resolution protocol to alternate docking and loop modeling steps. Use your code to model cetuximab. Use the job distributor to track your decoys. What is the lowest RMSD you can create in 100 decoys? _____

Workshop #9: Custom Movers & Energy Methods

You may have noticed that every "mover" we have introduced in the preceding chapters has shared an `apply()` method, which takes a pose as its argument. One of the benefits of object-oriented programming is the ability to design a group of objects, such as movers, that are all related to each other. More complicated objects are said to inherit methods and/or data from simpler objects, known as base classes. Rosetta 3 has been carefully designed in an organized way with countless derived classes inheriting code from higher up in the inheritance tree.

Suppose you wish to create your own movers or score function scoring components for a specific project. PyRosetta allows you to make your own custom movers and energy methods that inherit from Rosetta's `Mover` and `EnergyMethod` base classes just as movers like `SmallMover` and `MinMover` do. This workshop will provide a quick overview of classes and inheritance in Python and demonstrate how to customize movers and energy methods.

Suggested Reading

1. Python help on classes available at http://docs.python.org/tutorial/classes.html
2. A. Leaver-Fay, *et al.*, "ROSETTA3: An object-oriented software suite for the simulation and design of macromolecules," *Methods Enzymol.* **487**, 545–574 (2011).

Classes in Python

Python is more than just a scripting language; it is designed for object-oriented programming, including class definitions, inheritance, instantiation, class properties, and class methods. A full description of object-oriented programming in Python will not be covered here, but we will provide a quick overview, using a simple example.

Let's say we wish to create two objects in Python that contain properties and methods for drawing shapes on the screen — one for drawing a circle and another for drawing a square. Since a circle and a square are both shapes, they share a lot in common. It is a good programming practice to first create a "base class" `MyShape`, which will contain properties and methods that `MyCircle` and `MySquare` will "inherit".

The `class` statement is used to declare a class; the properties and methods for the class are defined indented underneath the class declaration in the class statement block:

```
class MyShape:
    """A base class for a generic shape."""
    def __init__(self):
        self.color = "black"  # default value for color
```

The `__init__()` method of any class — known as the "constructor" — is called whenever a new object is instantiated as part of that class. If we were to create a new `MyShape()` object using `shape = MyShape()`, the code below `def __init__(self):` would be run.

Note the use of the variable name `self` in the method declaration for `__init__()`. The keyword `self` refers to the particular instance of `MyShape` that is running the code. Whenever one calls *any* class method in Python, the first argument for the method is always the instance of the class calling the method. (For example, when we type `pose.total_residue()`, we are running a function within the `Pose` class and passing the variable `pose` as the first argument of `total_residue()`.) In the example above, the variable `color` stored within whichever object called `__init__()` is set to the value `"black"`. Such a variable stored within an object is called a "property".

Another common method for a class is `__str__()`. The function `__str__()` returns a string version of the object; that is, it will code for what happens if one tries to `print` the object:

```
def __str__(self):
    return self.__doc__
```

The keyword `__doc__` is a predefined variable in Python that stores the value of the "docstring" of its object. Any block comments after a declaration between a pair of three double quotes (`"""`) become the docstring for that class or method. If one were to type `print shape`, Python would return, in this case, `A base class for a generic shape.`

For our example class, we could code for a method that returns the area of the shape:

```
def area(self):
    """Return the area of the shape."""
    return # Code to calculate the area goes here.
```

Because `MyShape` is a base class and every shape has a different method for calculating its area, we simply return nothing here. This is a way to inform other programmers (or ourselves) that any shape objects created later should include and fully implement this method for that particular shape.

We could also include a method for outputting the shape on the screen. We will provide a default value for this method also. This allows one to call either `MyShape.draw()` or `MyShape.draw(<some_integer>)`, and either one will work.

```
def draw(self, line_width=1):
    """Draw the shape on the screen."""
    pass  # Code to draw the shape goes here.
```

Now we will code for a derived class, `MyCircle`. In our declaration line for `MyCircle`, we will include in parentheses the base class `MyShape`. This will cause `MyCircle` to "inherit" from `MyShape` all of its methods so we do not have to code them again. However, we will recode the `__init__()` and `area()` methods, because those are unique for a circle.

```
class MyCircle(MyShape):
    """A subclass of MyShape for a circle."""
    def __init__(self):
        # This overrides the __init__() method inherited
        # from MyShape.
        MyShape.__init__(self)
        self.radius = 1.0  # default value

    def area(self):
        """Return the area of the circle."""
        # This overrides the area() method inherited from
        # MyShape.
        return math.pi * self.radius**2
```

Notice how we call the `__init__()` method of the base class `MyShape`. Doing such will set the `color` property of the instance of `MyCircle` when it is initialized or constructed. In this way we can inherit a class's method and then add additional things to it, such as the addition of the definition of the property `radius` here.

One could make a similar class `MySquare` with different `__init__()` and `area()` methods.

1. Create a single Python file (`my_shapes.py`) containing the above classes. In addition, create a class `MySquare` that initializes a property `side_length` and includes a mathematically appropriate `area()` method. Import the two derived classes into IPython using `from my_shapes import MyCircle, MySquare`, and then run the following lines of code:

    ```
    circle = MyCircle()
    square = MySquare()
    print circle
    print square
    print circle.color
    print square.color
    square.side_length = 2
    print square.area()
    circle.radius, circle.color = 1.5, "pink"
    print circle.area()
    circle.draw(2)
    ```

    After typing the above lines, what is the area of your square? ____ What is the area of your circle? _____ Assuming that you had put actual code for drawing a circle into the `draw()` method, what would the color of the drawn circle be? _____ What would be the line width of the drawn circle? ___

Workshop #9: Custom Movers & Energy Methods | 63

Custom Mover Classes

Now we have a foundation for creating our own mover classes in PyRosetta that are subclasses of a base class `Mover`.

2. Create a Python file with the following lines of code. (Be sure to `import rosetta` at the top of the file, but you do not need to call `init()`.)

```
class PhiNByXDegreesMover(rosetta.protocols.moves.Mover):
    """A mover that increments the phi angle of residue N
    by X degrees.

    Default values are residue 1 and 15 degrees.

    """
    def __init__(self, N_in=1, X_in=15):
        """Construct PhiNByXDegreesMover."""
        rosetta.protocols.moves.Mover.__init__(self)

        self.N = N_in
        self.X = X_in

    def __str__(self):
        return "residue: " + str(self.N) + \
               "  phi increment: " + str(self.X) + \
               " degrees"
```

To function as a true Rosetta mover, we must do a few specific things. First, we must make our class inherit from `rosetta.protocols.moves.Mover`, a Rosetta base class. Furthermore, in the initialization code for our new mover object, we must run the `__init__()` constructor method of the Mover base class. One can also inherit from other Rosetta movers besides `Mover`. If this is done, simply call that particular mover's constructor instead. (You can — and should — include other mover initializations, such as is done in the code above, *e.g.*, `self.N = N_in`.)

Now expand the above class with the following two methods:

```
    def get_name(self):
        """Return name of class."""
        return self.__class__.__name__

    def apply(self, pose):
        """Applies move to pose."""
        pose = pose.get()
        print "Incrementing phi of residue", self.N, "by",
        print self.X, "degrees...."
        pose.set_phi(self.N, pose.phi(self.N) + self.X)
```

All movers must include the above two methods, namely, `get_name()` and `apply()`. The `get_name()` method for *every* PyRosetta mover you make need only include the line `return self.__class__.__name__`, which will return the name of the class of the object, in this case, `PhiNByXDegreesMover`.

The `apply()` method contains your custom code that alters the pose. Note that you must first call the `get()` method on `pose` as shown above. (The reason why this must be done is related to the underlying C++ access pointers and is unfortunately beyond the scope of this workshop.)

3. Import the `PhiNByXDegreesMover` class from the Python file containing it. Construct two instances of the mover, one with `N` set to 1 and `X` set to 15, the other with `N` set to 10 and `X` set to 45. (Note that properties of movers made in PyRosetta can be accessed and set directly — as you did for the properties in our shapes example — whereas, those from the Rosetta3 library must be accessed and set using getter and setter methods.) Load a test pose and apply both of your movers. Confirm that your movers behaved as expected using PyMOL.

4. Create a mover that decrements the psi angle of every *even* residue in a pose by a value passed during initialization. Bonus: See if you can do this by using a *single instance* of `PhiNByXDegreesMover` called from within a `for` loop in your new mover.

5. Create a `TrialMover` containing `PhiNByXDegreesMover` and one containing a similar `PsiNByXDegreesMover`. Write a script that runs a Monte Carlo algorithm using these `TrialMovers` to fold a small protein.

Decorators in Python

Setting up a custom energy method with a corresponding scoring component is more complicated than creating a new mover. Fortunately, to assist us with this process, we can use a "decorator", which helps us to prepare all of the required class methods.

Decorators are a high-level coding feature available in Python. A decorator is a function that takes a class, object, or method as input and returns a modified, or decorated, version of that class, object, or method. For example, suppose we have a class, `MyCircle`. We could define a decorator function such as the following:

```
def hollow(shape_in):
    """Modify the draw() method of an input shape class to
    output a hollow shape."""
    # Code to modify the draw() method goes here.
    return shape_out
```

We can now pass an instance (an object) of `MyCircle` to our decorator to get a modified version of the object:

```
circle = MyCircle()
circle.draw()    # Draws a filled circle.
hollow_circle = hollow(circle)
hollow_circle.draw()   # Draws a hollow circle.
```

We can also pass the class itself to our decorator function. In this example, we will overwrite our old `MyCircle` class with the new, decorated, hollow one:

```
MyCircle = hollow(MyCircle)
hollow_circle = MyCircle()
```

Python provides an alternate "wrapper" syntax, in which the entire class definition block of code is passed to the decorator function:

```
@hollow
class MyCircle(MyShape):
    # The rest of the class definition would go here.
```

If the `MyCircle` class definition is decorated like this by `@hollow`, then every `MyCircle` object will have a "hollow" `draw()` method.

Custom Energy Methods

It is unlikely that you will need to write your own decorator functions for structure prediction or design applications. However, there is a decorator in the PyRosetta library already written for you that will modify any custom energy method classes you write so that your custom classes will contain most of the required methods automatically. For example, here is how we can create a custom context-independent, one-body scoring method that will score a pose based solely on the number of residues. We will call our new class `LengthScoreMethod`. First, we must import the proper parent class method from Rosetta:

```
from rosetta.core.scoring.methods import
                    ContextIndependentOneBodyEnergy
```

Now, we will define our class:

```
@rosetta.EnergyMethod()
class LengthScoreMethod(ContextIndependentOneBodyEnergy):
    """A scoring method that favors longer peptides by
    assigning negative one Rosetta energy unit per
    residue.

    """
    def __init__(self):
        """Construct LengthScoreMethod."""
        ContextIndependentOneBodyEnergy.__init__(self,
                                        self.creator())

    def residue_energy(self, res, pose, emap):
        """Calculate energy of res of pose and set emap"""
        # 1 energy unit per residue
        emap.get().set(self.scoreType, -1.0)
```

And that's it! We should point out a few things. First, like when we inherited from `Mover`, we need to call the parent energy method class's `__init__()` constructor. An energy method constructor requires an additional argument, a "creator" function; where did this argument `self.creator()` come from? We did not explicitly define a method `creator()` to pass to `ContextIndependentOneBodyEnergy`'s constructor. The decorator, `EnergyMethod()`, (which is a callable class just like `ScoreFunction()`,) does that for us. `EnergyMethod()` also makes life easier by defining several other required methods for us.

Before we explain how this new energy method works, let's demonstrate how we can use it.

6. Create a Python file with the `LengthScoreMethod` code. (Include the proper imports.) Import Rosetta, import the new energy method, create a pose from the sequence "ACDEFGHIKLMNPQRSTVWY", and construct an empty `ScoreFunction()` with `sf = ScoreFunction()`. What is the score of your pose? ____

    Create a variable to hold the score type for your custom energy:

    ```
    len_score = LengthScoreMethod.scoreType
    ```

    (Note that we do not have to instantiate a `LengthScoreMethod` object; we simply can extract the score type directly from the class.)

    Now set the weight for the `len_score` scoring component to 1, just as we would for any other scoring component, such as `fa_atr`:

    ```
    sf.set_weight(len_score, 1.0)
    ```

    Score the pose. What is the score for the pose now? ____

Let's return to the code for our new energy method to understand how it works.

For one-body energy methods, whenever one scores a pose, the `ScoreFunction()` loops through each residue in the pose and calls the `residue_energy()` method of the energy method class associated with each score type. The `residue_energy()` method uses the residue and pose objects passed as arguments to calculate an energy score and sets the corresponding score type in the passed energy map. In our example above, we assign -1 energy unit per residue, independent of the context (*e.g.*, what type of residue it is), so we did not need to do anything with the arguments `res` or `pose`; we simply had to set `self.scoreType` in `emap` to -1.0.

Note that, as with custom mover `apply()` methods, because the `residue_energy()` method is called by a Rosetta function (written in C++, not a Python), we need to "`get()`" `emap` or any other passed object before we can use it.

7. Create an energy method that gives a favorable (-1.0) score for each aromatic residue in a pose. (Remember to call `res.get()`.) Repeat programming exercise 3c and d of Workshop 6, except add your aromatic-favoring energy method score type to the full atom standard score function to bias the design. How do your results change? How much do you have to adapt the weight of your score type before all of the designed residues are aromatic?

The procedure for making a custom two-body energy method is similar, but there are a few additional required methods and `residue_energy()` is replaced by `residue_pair_energy()`. Below is a template one can use for creating a context-independent, two body score type:

```
from rosetta.core.scoring.methods import
                        ContextIndependentTwoBodyEnergy


@rosetta.EnergyMethod()
class CI2BScoreMethod(ContextIndependentTwoBodyEnergy):
    """A scoring method that depends on pairs of residues.
    """
    def __init__(self):
        """Construct CI2BScoreMethod."""
        ContextIndependentTwoBodyEnergy.__init__(self,
                                    self.creator())

    def residue_pair_energy(self, res1, res2, pose, sf,
                            emap):
        """Calculate energy of each pair of res1 and res2
        of pose and set emap."""
        # A real method would calculate a value for score
        # from res1 and res2.
        score = 1.0
        emap.get().set(self.scoreType, score)

    def atomic_interaction_cutoff(self):
        """Get the cutoff."""
        return 0.0  # Change this value to set the cutoff.

    def defines_intrares_energy(self, weights):
        """Return True if intra-residue energy is
        Defined."""
        return True

    def eval_intrares_energy(self, res, pose, sf, emap):
        """Calculate intra-residue energy if defined."""
        pass
```

The template for a context-dependent, two-body method is identical, except that it inherits from and initializes `ContextDependentTwoBodyEnergy` instead.

Programming Exercises

1. Pick a small protein and use the `toolbox` method `get_secstruct(pose)` to determine the secondary structure, store that information in the pose, and output the results. Using the method `Pose.sectruct(resnum)`, and `PhiNByXDegreesMover`, loop through all residues that are a part of loops, set X between 1 and 15 stepping by 1

each time, apply for each case, and record the *change* in score. Reset the pose after each move. Plot the average change in score vs. X. Repeat this process for helix and strand residues, and then do the same three plots using `PsiNByXDegreesMover` instead. Do the sets of plots look similar for phi and psi? Does secondary structure appear to have an effect on how large X can be before the score is severely penalized? Do the default `angle_max` values of 0°, 5°, and 6° for helix, sheet, and loop, respectively, make sense based on your plots?

2. Create a context-dependent, two-body energy method to reward structures containing salt bridges. The energy method should give a bonus for each pair of acid and base side chains within 6 Å of each other. Find a small protein with several salt bridges. Run a *de novo* folding algorithm with and without your new scoring method. Does scoring for salt bridges give better results, that is, are the RMSDs from the native structure lower?

Coda

We hope these short tutorials have given you a broad set of basic abilities in protein structure prediction and design. PyRosetta's power is in its flexibility. You are now able to interactively combine folding, docking, and design operations and to use fold trees, move maps, and movers to tailor each operation to operate on the portion of the protein that is appropriate for the particular biological problem at hand.

There are many more features available in Rosetta and PyRosetta that you may find useful. Several particular useful features that are beyond the scope of this manual but currently implemented in PyRosetta include:

- Ligands composed of non-protein atoms or heteroatoms
- Nucleic acids, DNA and RNA
- Post-translationally modified and non-canonical amino acids
- Peptoids, carbohydrates, and other linear and branched polymers

Each of these can be loaded into poses, measured, scored, manipulated, designed or designed around, and docked. Ligands and modified amino acids require manipulation of the params files in the PyRosetta database. Example params files are provided in Appendix B. Future additions of this workbook will include workshops on these topics.

PyRosetta is being continually expanded, particularly with the expansion of the underlying Rosetta code. Please watch the website for future updates.

Appendix A: Command Reference

Python Commands and Syntax	

`# I am a comment; Python ignores me.`
`"""I am also ignored.`
`Me too!`
`"""`
	Comments
	Block comments (doc strings)

`i = 1  # (After the # is ignored.)`
`first_name = "Bob"`
`j = 1.0`
`i_am_a_boolean = True`
`i_am_an_integer = not i_am_a_boolean`
`k = first_name`
	Simple variable assignments (Python is case sensitive.)

`tuple = (1, 2, 3)`
`list = [15, 'X', -1.5, "lion"]`
`2_D_list[[1, 0], [0, 1]]`
`dict = {"apple": "sour", "days": 24}`
	Assignment to various sequences (tuples, lists, and dictionaries)

`print i + 2, 3/2, float(3)/2,` `        3.0/2.0, 2*j, (i + 3)**2`	Outputs `3 1 1.5 1.5 2.0 16` to the screen. (Python returns the floor of integer-only calculations, so convert at least one to a float if needed.)
`print k + " thinks " + str(i) +` `        " = 0."`	Outputs `Bob thinks 1 = 0.` (Python will not concatenate objects of different types, so a function must be used to convert an integer to a string.)

`print list[1], tuple[1:2], k[0:2],`
`        dict["days"],`
`        2_d_list[0][1]`
	Outputs `15 (2, 3) Bo 24 0`. (Lists, strings, *etc.*, are indexed starting with 0.)

`i += 1`
`j -= 1`
`print i, j`
	Increment and decrement operators.
	Outputs `2 0.0`

`tuple[2] = 4` `first_name[0] = 'R'`	Raise errors; tuples and strings are immutable.

`list[3] = "tiger"`
`dict["apple"] = "sweet"`
`list.append(False)`
`dict["new entry"] = 'Z'`
`print list, dict`
	Lists and dictionaries are mutable.
	
	Outputs `[15, 'X', -1.5, "tiger", False]`
	`{"apple": "sweet", "days": 24, "new entry", 'Z'}`

`for i in range(1, 10):` `    print i`	The newly defined variable `i` ranges from 1 up to *but not including* 10, and the command `print i` is executed for each value.

`for j in ("cats", "dogs", "fish"):`
`    print j`
	Outputs:
	`cats`
	`dogs`
	`fish`

`if x < 0:`
`    print "negative"`
`elif x == 0:`
`    print "zero"`
`else:`
`    print "positive"`
	Conditional statement that executes lines only if Boolean statements are true.
	`elif` means "or else, check if".
	Do not mix up = and ==!
	Use indenting to indicate blocks of code executed together under the conditional.

| ```python
if i_am_a_boolean:
    print "Yes!"
if not i_am_an_integer:
    print "No!"
``` | Outputs:<br>**Yes!**<br>**No!** |
| ```python
a, b = 1, 1
print a is b
print a == b
``` | Outputs:<br>**False**<br>**True**<br>(a and b contain the same value, but are not two names for the same object!) |
| ```python
while len(list) <= 7:
    list.append("blah")
print list
``` | Outputs:<br>[15, 'X', -1.5, "tiger", False, "blah", "blah", "blah"] |
| ```python
def my_func(a, b):
    c = a + b
    d = b + c
    e = c + d
    return c, d, e
``` | Defines a function (returns a value) or subroutine (does not). If a and b were 1 and 2, this function would return (3, 5, 8) |
| ```python
returned_values = my_func(1, 2)
value_of_c = returned_values[0]
value_of_d = returned_values[1]
value_of_e = returned_values[2]
``` | Syntax for using multiple values returned by a function, *e.g.*, value_of_c is 3. |
| ```python
class MyCircle:
    """This is a MyCircle class."""
    def __init__():
        #Code to run when a Circle
          object is instantiated
          goes here.
        self.radius = 1.0  # Sets
          default value for radius

    def draw(self, color=0):
        #Code to draw the circle
          goes here.
        pass
``` | Defines a class with two methods |
| ```python
circle = MyCircle()
circle.draw()
circle.radius = 1.5
circle.draw(1)
``` | Constructs a MyCircle object, draws a circle in color 0, and then draws a circle in color 1 |
| ```python
file = open("out.txt", 'w')
file.write("hello")
file.close()
``` | Opens a new file named out.txt for writing and outputs hello to the file. (Be sure to close your files when finished with them.) |
`import module`	Imports and runs the module module.py so that its functions can be called with module.function().
`from module import function`	Imports the specific function function so that it can be called with simply function().
`from module import *`	Imports all of the (public) functions from module.py.

Python Math

`math.exp(5)`	Returns the value of e^5
`math.pi`	Returns the value of π
`math.sin(theta)`	Returns the value of sin θ, where θ is in radians
`math.acos(x)`	Returns the value of arcos x in radians
`math.degrees(rad)`	Converts radians to degrees
`meth.radians(deg)`	Converts degrees to radians
`random.random()`	Returns a random floating point number between 0.0 and 1.0
`random.randint(5, 10)`	Returns a random integer between 5 and 10 (inclusive)
`random.gauss(10, 2)`	Returns a random number from a Gaussian distribution with a mean of 10 and a standard deviation of 2

Rosetta: Vector Calculus

`v = numeric.xyzVector_float(x, y, z)`	Creates a displacement vector with floating point precision from Cartesian coordinates
`print v` `print v.x, v.y, v.z`	Outputs `v` and its elements
`v - v2`	Returns the displacement vector between `v` and `v2`
`v.norm`	Returns the vector norm of `v`
`v.dot(v2)`	Returns the dot product of `v` and `v2`
`v.cross(v2)`	Returns the cross product of `v` and `v2`

Rosetta: Toolbox Methods

`cleanATOM("1YY8.pdb")`	Creates a "cleaned" pdb file with all non-`ATOM` lines of a pdb file removed
`cleanCRYS("1YY8.pdb", 2)`	Creates a "cleaned" crystal structure that removes redundant crystal contacts and isolates a monomer.
`pose = pose_from_rcsb("1YY8")`	Loads pdb `1YY8` from the Internet
`generate_resfile_from_pdb("input.pdb", "output.resfile")` `generate_resfile_from_pose(pose, "output.resfile")`	Generate a resfile from a pdb file or a pose, respectively
`mutate_residue(pose, 49, 'E')`	Replaces residue 49 of pose with a glutamate (E) residue (does not optimize rotamers)
`get_secstruct(pose)`	Assigns secondary structure information to `pose` and outputs it to the screen
`hbond_set = get_hbonds(pose)`	Instantiates and fills an H-bond set with hydrogen-bonding data

Rosetta: Pose Object	
`pose = Pose()`	Instantiates an empty `pose` object from the `Pose` class
`pose = pose_from_pdb("input_file.pdb")`	Loads a pdb file from the working directory into a new `pose` object
`pose = pose_from_rcsb("1YY8")`	Loads pdb 1YY8 from the Internet
`pose = pose_from_sequence("AAAAAA", "fa_standard")`	Creates a new pose from the given sequence using standard, full-atom residue type templates
`print pose`	Displays information about the pose object: pdb filename, sequence, and fold tree
`pose.sequence()`	Returns the sequence of the pose structure
`get_secstruct(pose)`	Assigns secondary structure information to `pose` and outputs it to the screen
`pose.assign(other_pose)`	Copies `other_pose` onto `pose`. You cannot simply type `pose = other_pose`, as that will only point `pose` to `other_pose` and not actually copy it.
`pose.dump_pdb("output_file.pdb")`	Creates a pdb file named `output_file.pdb` in the working directory using information from pose object.
`pose.is_fullatom()`	Returns `True` if the pose contains a full-atom representation of a structure
`pose.total_residue()`	Returns total number of residues in the pose

`pose.phi(5)`
`pose.psi(5)`
`pose.omega(5)`
`pose.chi(2, 5)`
	Returns the φ, ψ, ω, or χ_2 angles (in degrees) of the 5th residue in the pose

`pose.set_phi(5, 60.0)`
`pose.set_psi(5, 60.0)`
`pose.set_omega(5, 60.0)`
`pose.set_chi(2, 5, 60.0)`
	Sets the φ, ψ, ω, or χ_2 angles of the 5th residue in the pose to 60.0°

`print pose.residue(5)`	Outputs the amino acid details of residue 5
`pose.residue(5).name()`	Returns the 3-letter residue name for residue 5

`pose.residue(5).is_polar()`
`pose.residue(5).is_aromatic()`
`pose.residue(5).is_charged()`
	Return `True` if the 5th residue is of the queried type

`pose.residue(5).xyz("CA")` `pose.residue(5).xyz(2)`	Return the displacement vector of the α carbon (CA) of residue 5, which is the 2nd atom listed for that residue in the pose and a standard pdb file
`pose.residue(5).atom_index("CA")`	Returns 2
`for i in range (1, pose.total_residue() + 1):` `print pose.residue(i).name()`	Loops through all residues in the pose and outputs the 3-letter name of each (Unlike Python, Rosetta indexes residues starting with 1.)
`atom = pose.residue(5).atom("CA")`	Constructs an atom object for the α carbon (CA) of residue 5

`atom1 = AtomID(1, 5)`
`atom2 = AtomID(2, 5)`
`atom3 = AtomID(3, 5)`
	Construct unique atom *identifier* objects for the 1st, 2nd, and 3rd, atoms of *any* residue 5, respectively (This is not the same as the above command!)

`pose.conformation().bond_length(atom1, atom2)`	Returns the bond length (if stored in the `conformation` object) between `atom1` and `atom2`

`pose.conformation().bond_angle(atom1, atom2, atom3)`	Returns the bond angle in radians (if stored in the `conformation` object) of `atom1`, `atom2`, and `atom3`
`pose.conformation().set_bond_length( atom1, atom2, 1.5)`	Sets the bond length between atom1 and atom2 to 1.5 Å
`pose.conformation().set_bond_angle( atom1, atom2, atom3, 0.66666 * math.pi)`	Sets the bond angle of `atom1`, `atom2`, and `atom3` to ~120°
`print pose.pdb_info()`	Displays a table comparing the sequence numbering range in the pose with that of the pdb file from which the pose was generated
`pose.pdb_info().name()`	Returns the filename of the pdb file from which the pose was generated
`pose.pdb_info().number(5)`	Returns the pdb number of pose residue 5
`pose.pdb_info().chain(5)`	Returns the pdb chain label of pose residue 5
`pose.pdb_info().pdb2pose("A", 100)`	Returns which residue in the pose corresponds to residue 100 of chain A in the pdb file
`pose.pdb_info().pose2pdb(25)`	Returns a string containing the residue and chain label in the pdb file corresponding to residue 25 of the pose
`CA_rmsd(pose1, pose2)`	Returns the root-mean-squared deviation of the location of C_α atoms between the two poses

Rosetta Scoring Terms		
`fa_atr`	FA	van der Waals net attractive energy
`fa_rep`	FA	van der Waals net repulsive energy
`hbond_sr_bb, hbond_lr_bb`	FA/CEN	Hydrogen-bonding energies, short and long-range, backbone–backbone
`hbond_bb_sc, hbond_sc`	FA	Hydrogen-bonding energies, backbone–side-chain and side-chain–side-chain
`fa_sol`	FA	Solvation energies (Lazaridis–Karplus)
`fa_dun`	FA	Dunbrack rotamer probability
`fa_pair`	FA	Statistical residue–residue pair potential
`fa_intra_rep`	FA	Intraresidue repulsive Van der Waals energy
`fa_elec`	FA	Distance-dependent dielectric electrostatics
`pro_close`	FA	Proline ring closing energy
`dslf_ss_dst, dslf_cs_ang, dslf_ss_dih, dslf_ca_dih`	FA	Disulfide statistical energies (S–S distance, *etc.*)
`ref`	FA/CEN	Amino acid reference energy of unfolded state
`p_aa_pp`	FA/CEN	Propensity of amino acid in (φ,ψ) bin, $P(aa|\varphi,\psi)$
`rama`	FA/CEN	Ramachandran propensities
`vdw`	CEN	van der Waals "bumps" (repulsive only)
`env`	CEN	Residue environment score (statistical)
`pair`	CEN	Residue–residue pair score (statistical)
`cbeta`	CEN	β-carbon score

Rosetta: Scoring	
`scorefxn = ScoreFunction()`	Instantiates an empty `scorefxn` object from the `ScoreFunction` class
`scorefxn = get_fa_scorefxn()`	Constructs a score function with the default Rosetta full-atom energy terms and weights
`scorefxn = create_score_function("my_fxn")`	Constructs a score function with terms and weights from the `my_fxn` weights file
`scorefxn = create_score_function_ws_patch("my_fxn", "docking")`	Constructs a score function from the `my_fxn` weights file with a patch for docking simulations
`scorefxn.set_weight(fa_atr, 1.0)`	Sets the weight of the `fa_atr` term of the scoring function
`scorefxn.get_weight(fa_atr)`	Gets the weight of the `fa_atr` term of the scoring function
`print scorefxn`	Shows score function weights and details
`scorefxn(pose)`	Returns the score of `pose` with the defined function `scorefxn` and stores the results in the `energies` object within `pose`
`scorefxn.show(pose)`	Returns a table of weights and raw & weighted scores broken down by scoring term
`pose.energies().show()`	Shows the breakdown of all energies (except backbone hydrogen-bonding energies) in the pose by residue
`pose.energies().show(5)`	Shows the breakdown of all energy contributions (except backbone hydrogen-bonding energies) from residue 5
`etable_atom_pair_energies(atom1, atom2, scorefxn)`	Returns a tuple of the attractive, repulsive, and solvation score components of a pair of Atom obects (not `AtomID`s!)
`pose.energies().total_energies()[fa_atr]`	Returns the `fa_atr` contribution to the total energy
`pose.residue_total_energies(5)[fa_atr]`	Returns the `fa_atr` contributions from residue 5
`hbond_set = hbonds.HBondSet()`	Instantiates an empty set for storing hydrogen-bonding energies and information
`pose.update_residue_neighbors()` `hbonds.fill_hbond_set(pose, False, hbond_set)`	Updates the `Energies` object within `pose` based on neighboring residues and fills `hbond_set` with this data (The option `False` is to forgo calculating a derivative.)
`hbond_set = get_hbonds(pose)`	Combines the steps above to instantiate and fill a set with hydrogen-bonding data
`hbond_set.show(pose)`	Shows a listing of all hydrogen bonds and their energies in a given pose
`hbond_set.show(pose, 5)`	Shows a listing of the hydrogen bonds and their energies from residue 5 of `pose`
`emap = EMapVector()`	Instantiates an energy map object to store a vector of scores
`scorefxn.eval_ci_2b(5, 6, pose, emap)`	Evaluates context-independent two-body energies between pose residues 5 and 6 and stores the energies in the energy map
`emap[fa_atr]`	Returns the `fa_atr` term from the map

Appendix A: Command Reference

PyMOL Mover	
`pmm = PyMOL_Mover()`	Instantiates the PyMOL mover
`pmm.apply(pose)`	Sends the pose coordinates to PyMOL for viewing
`pmm.send_energy(pose)`	Instructs PyMOL to color the pose by its total energy
`pmm.send_energy(pose, label=True)`	Instructs PyMOL to color the pose by its total energy and label each Cα with the value.
`pmm.send_energy(pose, "fa_atr")`	Instructs PyMOL to color the pose by its fa_atr contribution
`pmm.label_energy(pose)`	Instructs PyMOL to label each Cα with the value of its total energy contribution
`pmm.energy_type(fa_atr)` `pmm.update_energy(True)`	Sets the PyMOL mover to color by fa_atr every time the pose is updated with `pmm.apply(pose)`
`pmm.keep_history(True)`	Instructs PyMOL to store all pose conformations in separate frames
`colors = {1:"red", 2:"blue"}` `pmm.send_colors(pose, colors, "gray")`	Instructs PyMOL to color residue 1 red, 2 blue, and all others gray
`pmm.send_hbonds(pose)`	Instructs PyMOL to display distance lines for every hydrogen bond
`pmm.send_ss(pose)`	Uses DSSP to reassign secondary-structure and instructs PyMOL to display it as a cartoon
`pmm.send_polars(pose)`	Instructs PyMOL to color polar residues red and nonpolar residues blue
`pmm.send_movemap(pose, mm)`	Instructs PyMOL to color movable regions of the pose green and non-movable regions red
`pmm.send_foldtree(pose)`	Instructs PyMOL to color cutpoints red, jump points orange, and loop regions a unique color
`pmm.view_foldtree_diagram(pose)`	Draws a 3-D fold tree diagram in PyMOL
`observer = PyMOL_Observer(pose, True)`	Updates PyMOL anytime a change is made to `pose` and keeps a history

Residue Type Set Movers	
`switch = SwitchResidueTypeSetMover("centroid")`	Instantiates a mover object that will change poses to the centroid residue type set (`"fa_standard"` is also available.)
`switch.apply(pose)`	Changes `pose` to the centroid residue type set
`recover_sidechains = ReturnSidechainMover(initial_fa_pose)`	Instantiates a mover object that will return the side chains and rotamers from an initial full-atom pose to a centroid version of the same peptide
`recover_sidechains.apply(pose)`	Changes `pose` to a full-atom type set and sets the rotamers to those found in `initial_fa_pose`

MoveMap

`movemap = MoveMap()`	Instantiates a `movemap` object from the `MoveMap` class
`movemap.show(5)`	Displays the movemap settings for residues 1 to 5
`movemap.set_bb(True)`	Allows all backbone torsion angles to vary
`movemap.set_chi(True)`	Allows all side-chain torsion angles (χ) to vary
`movemap.set_bb(10, False)` `movemap.set_chi(10, False)`	Forbid the backbone and side-chain torsion angles of residue 10 from varying
`movemap.set_bb_true_range(10, 20)` `movemap.set_chi_true_range(10, 20)`	Allow backbone and side-chain torsion angles to vary in residues 10 to 20, inclusive
`movemap.set_jump(1, True)`	Allows jump #1 to be flexible

Fragment Movers

`fragset = ConstantLengthFragSet(3)`	Constructs a 3-mer fragment set object
`fragset.read_fragment_file("fragfile")`	Loads data from the file `fragfile` into `fragset` (A fragment file must be downloaded from the Robetta server.)
`mover_3mer = ClassicFragmentMover(fragset, movemap)`	Constructs a fragment mover using the fragment set and the movemap
`mover_3mer.apply(pose)`	Replaces the angles in `pose` with those from a random 3-mer fragment from `fragset`, only in positions allowed by `movemap`
`smoothmover = SmoothFragmentMover(fragset, movemap)`	Constructs a "smooth" fragment mover (Fragment "insertions" are followed by a second, downstream fragment insertion chosen to minimize global disruption.)

Small and Shear Movers

`kT = 1.0`	Variable simulating the product of the Boltzmann constant and temperature (1.0 approximates room temperature.)
`smallmover = SmallMover(movemap, kT, 5)` `shearmover = ShearMover(movemap, kT, 5)`	Constuct a small or shear mover with a movemap, a temperature, and 5 moves
`smallmover.angle_max("H", 15)` `shearmover.angle_max("H", 15)`	Set the maximum change in dihedral angle within helix residues to 15° (`"E"` sets the maximum for sheet residues; `"L"` loop residues.)
`smallmover.apply(pose)` `shearmover.apply(pose)`	Apply the movers

Appendix A: Command Reference

Minimize Mover

`minmover = MinMover()`	Consructs a minimize mover with default arguments
`minmover = MinMover(movemap, scorefxn, "linmin", 0.01, True)`	Construct a steepest descent minimize mover with a particular `MoveMap` and `ScoreFunction` and a score tolerance of 0.01
`minmover.movemap(movemap)`	Sets a movemap
`minmover.score_function(scorefxn)`	Sets a score function
`minmover.min_type("linmin")`	Sets a the minimization type to a line minimization (one direction in space), *i.e.*, "steepest descent"
`minmover.min_type("dfpmin")`	Sets a the minimization type to a David–Fletcher–Powell minimization (multiple iterations of `"linmin"` in conjugate directions)
`minmover.tolerance(0.5)`	Sets the mover to iterate until within 0.5 score points of the minimum
`minmover.apply(pose)`	Minimizes the pose

Monte Carlo Object

`mc = MonteCarlo(pose, scorefxn, kT)`	Constructs a `MonteCarlo` object for a given pose and score function at a temperature of `kT`
`mc.set_temperature(1.0)`	Sets the temperature in the `MonteCarlo` object
`mc.boltzmann(pose)`	Accepts or rejects the current pose, compared to the last pose, according to the standard Metropolis criterion
`mc.show_scores()`	Shows stored scores, counts of moves accepted/rejected, or both, respectively.
`mc.show_counters()`	
`mc.show_state()`	
`mc.recover_low(pose)`	Sets the pose to the lowest-energy configuration ever encountered during the search
`mc.reset(new_pose)`	Resets all counters and sets the lowest and last pose stored to `new_pose`.

Trial Mover

`smalltrial = TrialMover(smallmover, mc)`	Constructs a combination mover that will apply the small mover, then call the `MonteCarlo` object `mc` to accept or reject the new pose
`smalltrial.num_accepts()`	Returns the number of times the move was accepted
`smalltrial.acceptance_rate()`	Returns the acceptance rate of the moves

Sequence Movers and Repeat Movers

`seqmover = SequenceMover()`	Construct a combination mover that will call a series of other movers in sequence
`seqmover.addmover(smallmover)`	
`seqmover.addmover(shearmover)`	
`seqmover.addmover(minmover)`	
`repeatmover = RepeatMover(fragmover, 10)`	Constructs a combination mover that will call `fragmover` 10 times
`randommover = RandomMover()`	Construct a combination mover that will randomly apply one of a set of movers each time it is applied
`randmover.addmover(smallmover)`	
`randmover.addmover(shearmover)`	
`randmover.addmover(minmover)`	

Classic Relax Protocol

`relax = ClassicRelax()`	Instantiates an object that encompasses the entire standard Rosetta refinement protocol as presented in Bradley, Misura, & Baker 2005
`relax.set_scorefxn(scorefxn)`	Sets the score function
`relax.apply(pose)`	Applies the whole protocol

Packer Task Object

`task = standard_packer_task(pose)`	Constructs a packer task `object` with instructions to repack all residues in `pose` using default rotamer library options, without repacking disulfide bonds
`task = TaskFactory.create_packer_task(pose)`	Constructs a default packer task `object` without any extra rotamer options
`task.restrict_to_repacking()`	Restricts all residues to repacking only (no design/"mutations")
`task.temporarily_fix_everything()`	Fixes/locks all residues' rotamers (no repacking)
`task.temporarily_set_pack_residue(5, True)`	Sets residue 5 to allow repacking
`generate_resfile_from_pdb("input.pdb", "output.resfile")`	Generate a resfile from a pdb file or a pose, respectively
`generate_resfile_from_pose(pose, "output.resfile")`	
`parse_resfile(pose, task, "file.resfile")`	Sets packer task for `pose` based on instructions in resfile

Resfile Codes

NATRO	Use the na̱ti̱ve amino acid residue and na̱ti̱ve ro̱tamer (do not repack)
NATAA	Use the na̱ti̱ve a̱mino a̱cid residue but allow repacking to other rotamers
PIKAA ILV	Pi̱ck from a̱mino a̱cid residues Ile, Leu, and Val and allow repacking
ALLAA	Use a̱ll a̱mino a̱cid residues and allow repacking

Side Chain Packing Mover

`pack_mover = PackRotamersMover(scorefxn, task)`	Constructs a mover that will use instructions from the packer task to optimize or "mutate" side chain conformations in the pose

Simple Point Mutation

`mutate_residue(pose, 49, 'E')`	Replaces residue 49 of pose with a glutamate (E) residue (does not optimize rotamers)

Fold Tree

`ft = FoldTree()`	Constructs an empty fold tree
`ft = pose.fold_tree()`	Extracts the current fold tree from the `pose`
`pose.fold_tree(ft)`	Attaches the fold tree `ft` to the pose
`ft.add_edge(1, 30, -1)`	Creates a peptide edge (code -1) from residues 1 to 30 (This edge will build N-to-C)
`ft.add_edge(100, 31, -1)`	Creates a peptide edge from residues 100 to 31 (This edge will build C-to-N.)
`ft.add_edge(30, 100, 1)`	Creates a jump (rigid-body connection) between residues 30 and 100 (The jump number is 1; each jump needs a unique, sequential jump number.)
`ft.check_fold_tree()`	Returns `True` only for valid trees
`print ft`	Prints the fold tree
`ft.simple_tree(100)`	Creates a single-peptide-edge tree for a 100-residue protein
`ft.new_jump(40, 60, 50)`	Creates a jump from residue 40 to 60, a cutpoint between 50 and 51, and splits up the original edges to finish the tree
`ft.clear()`	Deletes all edges in the fold tree
`setup_foldtree(pose, "A_B", Vector1([1]))`	Creates a fold tree for `pose` with jump #1 between the centers of mass of chains A and B
`set_single_loop_fold_tree(pose, loop)`	Creates a fold tree for `pose` with jump points and a cutpoint defined by a Loop object, and splits up the original edges to finish the tree (See below for the `Loop` object.)

Jump Object

`pose.jump(1).get_rotation()`	Returns the rotation matrix for the jump
`pose.jump(1).get_translation()`	Returns the translation vector for the jump

Rigid Body Movers

`pert_mover = RigidBodyPerturbMover(1, 8, 3)`	Constructs a mover that will make a random rigid-body move of the downstream partner across jump #1 (Rotations and translations are chosen from a Gaussian with a mean of 8° and 3 Å, respectively.)

`trans_mover = RigidBodyTransMover(pose, jump_num)`
`trans_mover.trans_axis(a)`
`trans_mover.step_size(5)`
	Constructs a mover that will translate two partners, defined by jump_num, along an axis defined by vector a by 5 Å

`spin_mover = RigidBodySpinMover(jump_num)`
`spin_mover.spin_axis(axis)`
`spin_mover.rot_center(center)`
`spin_mover.angle_size(45)`
	Constructs a mover that will spin the chain downstream of jump_num around a spin axis and rotation center by 45° (No specified angle size randomizes the spin.)

`random_mover = RigidBodyRandomizeMover(pose, 1, partner_upstream)` `random_mover = RigidBodyRandomizeMover(pose, 1, partner_downstream)`	Construct a mover that will rotate one of the partners across jump #1 randomly about its geometric center (`partner_upstream` and `partner_downstream` are predefined constants, not variables.)
`slide = DockingSlideIntoContact(1)` `slide = FADockingSlideIntoContact(1)`	Construct movers to translate two centroid or full-atom chains across jump #1 into contact, respectively

Appendix A: Command Reference

Docking Protocols

`dock_lowres = DockingLowRes(scorefxn_low, jump_num)`	Constructs a low-resolution, centroid-based Monte Carlo search protocol (50 rigid-body perturbations with adaptable step sizes)
`dock_hires = DockMCMProtocol(scorefxn_high, jump_num)`	Constructs a high-resolution, full-atom-based Monte Carlo search protocol with rigid-body moves, side-chain packing, and minimization

Loop Objects

`loop = Loop(15, 24, 20)`	Defines a loop with stems at residues 15 and 24, and a cutpoint at residue 20
`loops = Loops()`	Constructs an object to contain a set of loops
`loops.add_loop(loop1)`	Adds a `Loop` object to `loops`

Loop Movers

`ccd = CcdLoopClosureMover(loop1, movemap)`	Creates a mover which performs Canutescu & Dunbrack's cyclic coordinate descent loop closure algorithm
`loop_refine = LoopMover_Refine_CCD(loops)`	Creates a high-resolution refinement protocol consisting of cycles of small and shear moves, side-chain packing, CCD loop closure, and minimization.

RMSD-Calculating Functions

`CA_rmsd(pose1, pose2)`	Returns the RMSD between the Cα atoms of `pose1` and `pose2`
`calc_Lrmsd(pose1, pose2, Vector([1]))`	Return the ligand RMSD between `pose1` and `pose2`
`loop_rmsd(pose, ref_pose, loops, True)`	Returns the RMSD of all loops in the reference frame of the fixed protein structure

Job Distributor

`jd = PyJobDistributor("output", 10, scorefxn)`	Constructs a job distributor that will create 10 model structures named output_1.pdb to output_9.pdb and a file containing a table of scores
`jd.native_pose = native_pose`	Sets the native pose for RMSD comparisons
`jd.job_complete`	Returns True if all decoys have been output
`jd.output_decoy(pose)`	Outputs `pose` to a file and increments the decoy number

`while not jd.job_complete:`
`    # Code for creating decoys`
`    jd.output_decoy(pose)`
	Loop to create decoys until all have been output

`jd.additional_decoy_info = "Created by Andy"`	Sets a string to be output to the pdb files

Appendix B: Residue Parameter Files

Parameter files describing the chemical and structural properties of each residue is found in the PyRosetta package in the `rosetta_database/chemical/residue_type_sets` directory.

The full-atom residue parameters are stored in the `/fa_standard/residue_types` directory. As an example, the parameter file for threonine is shown below.

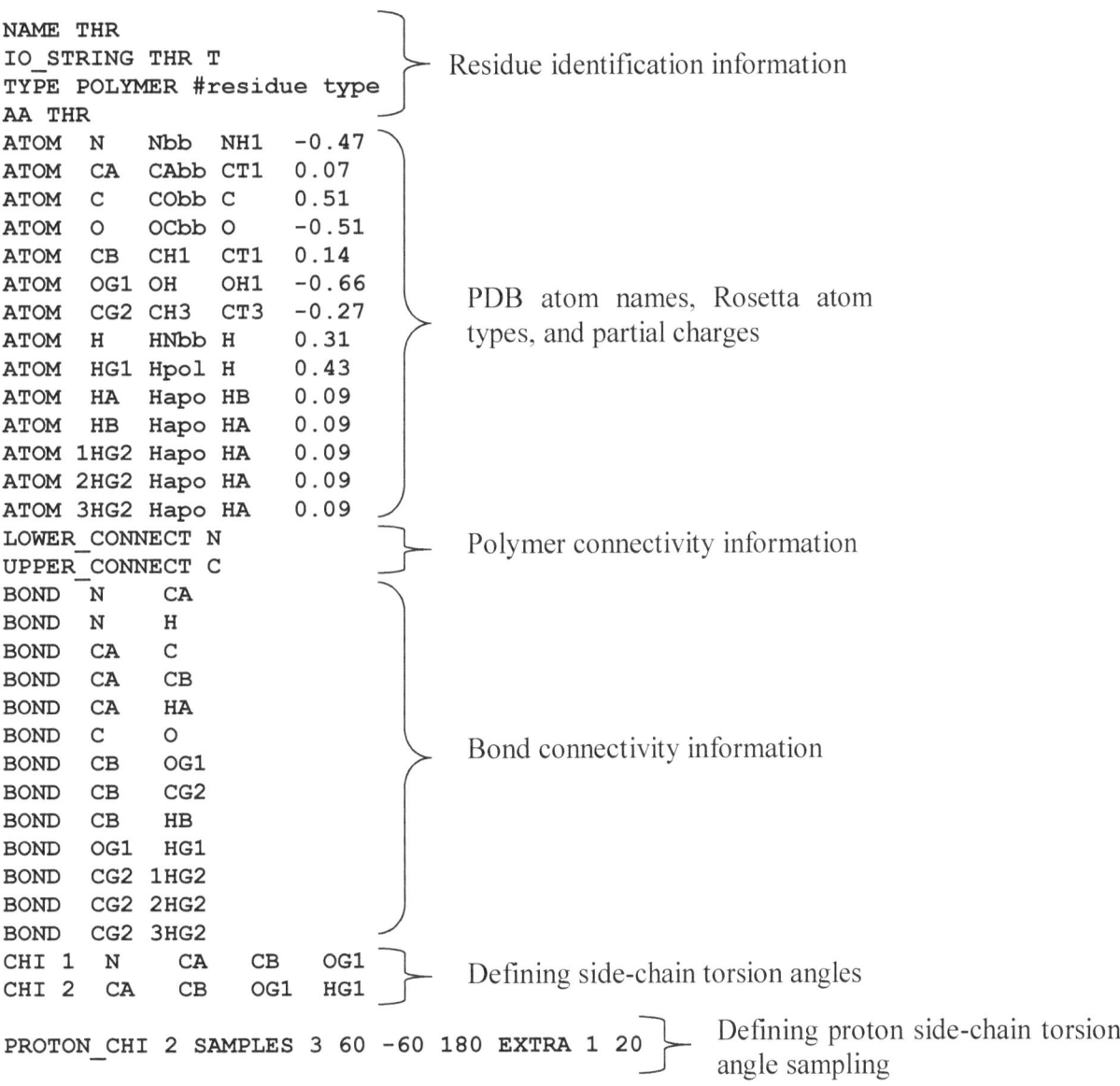

```
NAME THR
IO_STRING THR T
TYPE POLYMER  #residue type
AA THR
```
⎫ Residue identification information

```
ATOM  N    Nbb   NH1  -0.47
ATOM  CA   CAbb  CT1   0.07
ATOM  C    CObb  C     0.51
ATOM  O    OCbb  O    -0.51
ATOM  CB   CH1   CT1   0.14
ATOM  OG1  OH    OH1  -0.66
ATOM  CG2  CH3   CT3  -0.27
ATOM  H    HNbb  H     0.31
ATOM  HG1  Hpol  H     0.43
ATOM  HA   Hapo  HB    0.09
ATOM  HB   Hapo  HA    0.09
ATOM  1HG2 Hapo  HA    0.09
ATOM  2HG2 Hapo  HA    0.09
ATOM  3HG2 Hapo  HA    0.09
```
⎫ PDB atom names, Rosetta atom types, and partial charges

```
LOWER_CONNECT N
UPPER_CONNECT C
```
⎫ Polymer connectivity information

```
BOND  N    CA
BOND  N    H
BOND  CA   C
BOND  CA   CB
BOND  CA   HA
BOND  C    O
BOND  CB   OG1
BOND  CB   CG2
BOND  CB   HB
BOND  OG1  HG1
BOND  CG2  1HG2
BOND  CG2  2HG2
BOND  CG2  3HG2
```
⎫ Bond connectivity information

```
CHI 1  N   CA   CB   OG1
CHI 2  CA  CB   OG1  HG1
```
⎫ Defining side-chain torsion angles

```
PROTON_CHI 2 SAMPLES 3 60 -60 180 EXTRA 1 20
```
⎫ Defining proton side-chain torsion angle sampling

```
PROPERTIES PROTEIN POLAR          ⎫— Residue properties
NBR_ATOM CB                       ⎫
NBR_RADIUS 3.4473                 ⎬— Defining parameters for neighbor
FIRST_SIDECHAIN_ATOM CB           ⎭   calculations
ACT_COORD_ATOMS OG1 END
ICOOR_INTERNAL    N         0.000000    0.000000   0.000000   N    CA   C
ICOOR_INTERNAL    CA        0.000000  180.000000   1.458001   N    CA   C
ICOOR_INTERNAL    C         0.000000   68.800049   1.523257   CA   N    C
ICOOR_INTERNAL    UPPER   149.999954   63.800026   1.328685   C    CA   N
ICOOR_INTERNAL    O       180.000000   59.199905   1.231016   C    CA   UPPER
ICOOR_INTERNAL    CB     -121.983574   68.467087   1.539922   CA   N    C
ICOOR_INTERNAL    OG1      -0.000077   70.419235   1.433545   CB   CA   N
ICOOR_INTERNAL    HG1       0.000034   70.573135   0.960297   OG1  CB   CA
ICOOR_INTERNAL    CG2    -120.544136   69.469185   1.520992   CB   CA   OG1
ICOOR_INTERNAL    1HG2   -179.978256   70.557961   1.089826   CG2  CB   CA
ICOOR_INTERNAL    2HG2    120.032188   70.525108   1.089862   CG2  CB   1HG2
ICOOR_INTERNAL    3HG2    119.987984   70.541740   1.089241   CG2  CB   2HG2
ICOOR_INTERNAL    HB     -120.292923   71.020676   1.089822   CB   CA   CG2
ICOOR_INTERNAL    HA     -120.513664   70.221680   1.090258   CA   N    CB
ICOOR_INTERNAL    LOWER  -149.999969   58.300030   1.328684   N    CA   C
ICOOR_INTERNAL    H       180.000000   60.849979   1.010000   N    CA   LOWER
```

Residue structure defined in internal coordinates

Appendix B: Residue Parameter Files

The centroid residue parameters can be found in the /centroid/residue_types directory. The centroid parameter file for Threonine is shown below.

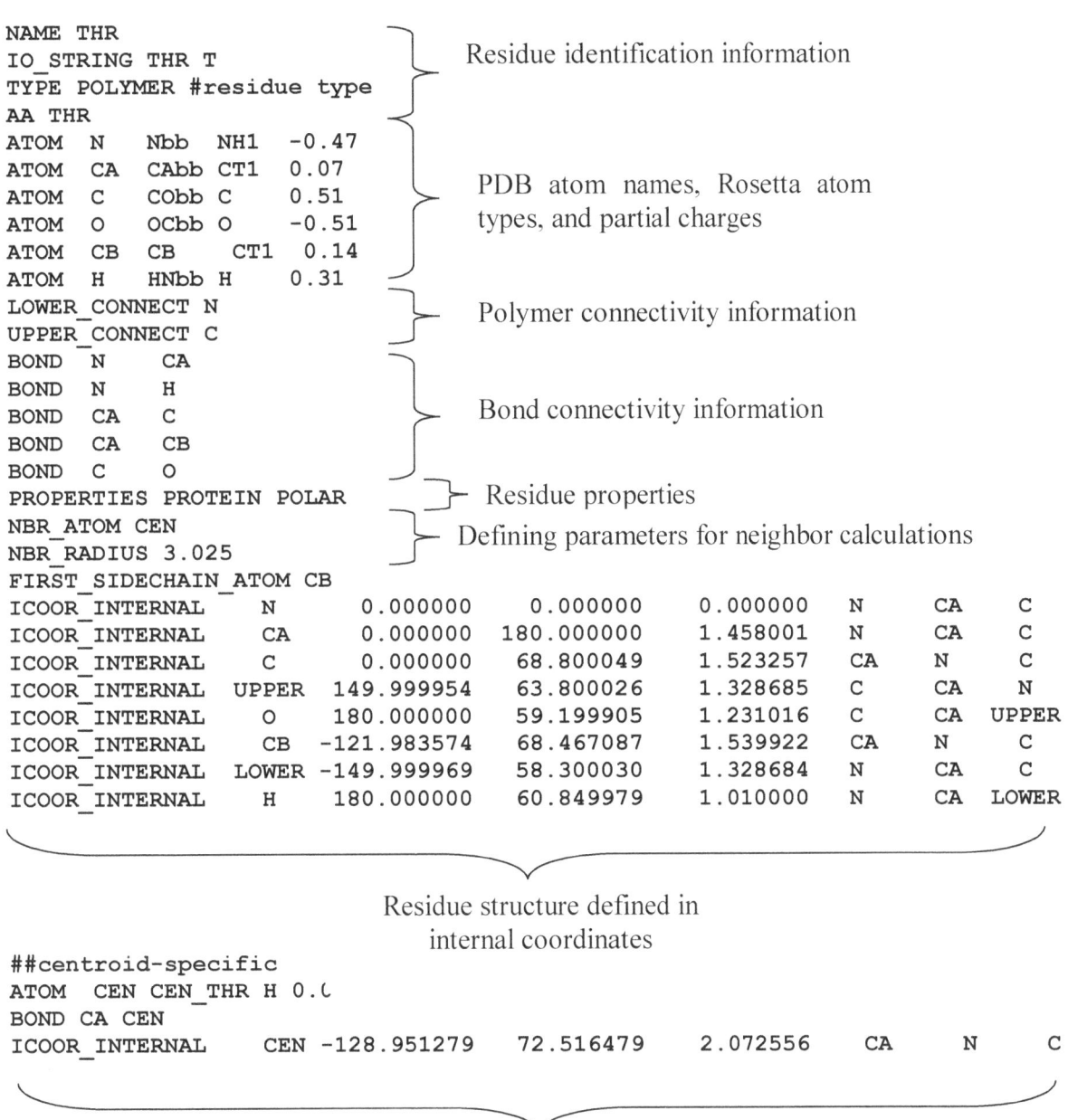

```
NAME THR
IO_STRING THR T                            ⎫
TYPE POLYMER #residue type                 ⎬  Residue identification information
AA THR                                     ⎭
ATOM  N   Nbb   NH1  -0.47    ⎫
ATOM  CA  CAbb  CT1   0.07    ⎪
ATOM  C   CObb  C     0.51    ⎬  PDB atom names, Rosetta atom
ATOM  O   OCbb  O    -0.51    ⎪  types, and partial charges
ATOM  CB  CB    CT1   0.14    ⎪
ATOM  H   HNbb  H     0.31    ⎭
LOWER_CONNECT N               ⎫
                              ⎬  Polymer connectivity information
UPPER_CONNECT C               ⎭
BOND  N   CA                  ⎫
BOND  N   H                   ⎪
BOND  CA  C                   ⎬  Bond connectivity information
BOND  CA  CB                  ⎪
BOND  C   O                   ⎭
PROPERTIES PROTEIN POLAR         ⎬ Residue properties
NBR_ATOM CEN                     ⎬ Defining parameters for neighbor calculations
NBR_RADIUS 3.025
FIRST_SIDECHAIN_ATOM CB
ICOOR_INTERNAL    N      0.000000    0.000000   0.000000   N   CA   C
ICOOR_INTERNAL    CA     0.000000  180.000000   1.458001   N   CA   C
ICOOR_INTERNAL    C      0.000000   68.800049   1.523257   CA  N    C
ICOOR_INTERNAL    UPPER 149.999954  63.800026   1.328685   C   CA   N
ICOOR_INTERNAL    O    180.000000   59.199905   1.231016   C   CA   UPPER
ICOOR_INTERNAL    CB  -121.983574   68.467087   1.539922   CA  N    C
ICOOR_INTERNAL    LOWER -149.999969 58.300030   1.328684   N   CA   C
ICOOR_INTERNAL    H    180.000000   60.849979   1.010000   N   CA   LOWER
```

Residue structure defined in internal coordinates

```
##centroid-specific
ATOM  CEN CEN_THR H 0.0
BOND CA CEN
ICOOR_INTERNAL    CEN  -128.951279   72.516479   2.072556   CA   N   C
```

Centroid-specific information

Appendix C: Cleaning pdb Files

Many pdb files have extraneous information and often do not conform to file standards. You may have to "clean" your pdb file before loading it into PyRosetta. You can do this through a variety of methods.

- From within a UNIX shell:

    ```
    grep "^ATOM" 1ABC.pdb > 1ABC.clean.pdb
    ```

- From within a DOS shell:

    ```
    findstr /b "ATOM" 1ABC.pdb > 1ABC.clean.pdb
    ```

- Using a PyRosetta `toolbox` method `cleanATOM`

    ```
    from toolbox import cleanATOM
    cleanATOM("1YY8.pdb")
    ```

    In this example, the `toolbox` method `cleanATOM` will create a file called `1YY8.clean.pdb`, with all non-`ATOM` lines removed. Warning: this method will overwrite any other files of the same name in its directory.

All of these methods will remove all lines that do not begin with `ATOM` in the pdb file and create a new "clean" pdb file named `1ABC.clean.pdb`.

One could also easily write a script to clean multiple pdb files at once. Here is an example Bash (UNIX shell) script that will clean all pdb files in a single directory:

```
#!/bin/sh
for pdbfile in *.pdb
do
        echo "cleaning $pdbfile..."
        grep ^ATOM $pdbfile > ${pdbfile%.pdb}.clean.pdb
done
```

If after cleaning a pdb file, PyRosetta still gives errors, you might use your text editor to open the pdb file and edit or remove the offending data lines manually.

Appendix D: Links to Online Help

PyMOL

- http://www.pymol.org/support — support
- http://www.pymolwiki.org — wiki
- http://www.pyrosetta.org/pymol_mover-tutorial — PyMOL_Mover tutorial

Python

- http://docs.python.org — documentation
- http://wiki.python.org — wiki
- http://www.python.org/dev/peps/pep-0008 — style guide for Python code

PyRosetta

- http://www.pyrosetta.org/tutorials — these workshops
- http://www.pyrosetta.org/scripts — sample scripts and useful tools
- http://www.pyrosetta.org/documentation — documentation

Rosetta

- http://www.rosettacommons.org/support — C++ documentation
- http://www.rosettacommons.org/forum — forum

www.ingramcontent.com/pod-product-compliance
Lightning Source LLC
Chambersburg PA
CBHW081142170526
45165CB00008B/2768